Geologic Controls on Radon

Edited by

Alexander E. Gates
Department of Geology
Rutgers University
Newark, New Jersey 07102

and

Linda C. S. Gundersen
U.S. Geological Survey
Box 25046, Federal Center, MS-939
Denver, Colorado 80225

271

1992

© 1992 The Geological Society of America, Inc.
All rights reserved.

All materials subject to this copyright and included
in this volume may be photocopied for the noncommercial
purpose of scientific or educational advancement.

Copyright is not claimed on any material prepared
by government employees within the scope of their
employment.

Published by The Geological Society of America, Inc.
3300 Penrose Place, P.O. Box 9140, Boulder, Colorado 80301

Printed in U.S.A.

GSA Books Science Editor Richard A. Hoppin

Library of Congress Cataloging-in-Publication Data
Geologic controls on radon / edited by Alexander E. Gates and Linda
   C. S. Gundersen.
     p. cm. — (Special paper ; 271)
    Includes bibliographical references.
    ISBN 0-8137-2271-3
    1. Radon. 2. Geochemistry. I. Gundersen, L. C. S. II. Series:
Special papers (Geological Society of America) ; 271.
QE516.R6G46   1992
553.4'97—dc20   92-32393
                                      CIP

Cover: The atomic symbol for radon.

10 9 8 7 6 5 4 3 2 1

Contents

Preface v

1. Geology of Radon in the United States 1
   Linda C. S. Gundersen, R. Randall Schumann, James K. Otton,
   Russell F. Dubiel, Douglass E. Owen, and Kendell A. Dickinson

*2. Sensitivity of Soil Radon to Geology and the Distribution of Radon
   and Uranium in the Hylas Zone Area, Virginia* 17
   Alexander E. Gates and Linda C. S. Gundersen

*3. Geologic and Environmental Implications of High Soil-Gas Radon
   Concentrations in the Great Valley, Jefferson and Berkeley Counties,
   West Virginia* 29
   Art Schultz, Calvin R. Wiggs, and Stephen D. Brower

*4. Soil Radon Distribution in Glaciated Areas: An Example from the
   New Jersey Highlands* 45
   Alexander E. Gates, Lawrence Malizzi, and John Driscoll III

5. Radon in the Coastal Plain of Texas, Alabama, and New Jersey 53
   Linda C. S. Gundersen and R. Thomas Peake

*6. Effects of Weather and Soil Characteristics on Temporal Variations
   in Soil-Gas Radon Concentrations* 65
   R. Randall Schumann, Douglass E. Owen, and Sigrid Asher-Bolinder

7. A Theoretical Model for the Flux of Radon from Rock to Ground Water .. 73
   Richard B. Wanty, Errol P. Lawrence, and Linda C. S. Gundersen

*8. The Influence of Season, Bedrock, Overburden, and House
   Construction on Airborne Levels of Radon in Maine Homes* 79
   E. Melanie Lanctot, Peter W. Rand, Eleanor H. Lacombe,
   C. Thomas Hess, and Gregory F. Bogdan

Preface

Surveys of the distribution of radon concentrations in soil-gas have been conducted for uranium exploration since at least the late 1950s. Because radon is the daughter of uranium (^{238}U), relatively high concentrations of radon commonly indicate the presence of uranium ore bodies. Because radon is the only gaseous daughter of uranium, as well as a noble gas, it can migrate much more easily than any of the other daughters. Therefore, radon can be used to evaluate as large and as deep an area as possible with the fewest samples. These evaluations yield quick results and have been credited with locating ore bodies up to 200 m deep.

Coincident with the decline of uranium exploration, environmental radon became an issue of concern. For example, uranium miners exposed to high concentrations of radon contracted lung cancer at a much higher than average rate. When concentrations of radon in homes exceeded those in uranium mines, radon went from obscurity to one of the most feared environmental hazards. The U.S. Environmental Protection Agency estimates that 15,000 to 25,000 deaths result from radon-induced lung cancer each year in the United States.

Radon has been studied more by health physicists than by any other group. Their findings, however, are constrained by the geologic environment because of the three processes responsible for indoor radon. The first two processes, generation and transmission, are directly controlled by geology. Concentration, the third process, is less so.

This volume shows the geologic controls on radon in terms of its distribution, processes, and limitations on how it can be used as a tool for geologic interpretation. The first chapter discusses the distribution of indoor radon across the United States and the geologic controls on that distribution. The next two chapters are process-oriented studies showing a new process to concentrate radon in areas with little uranium and radon's use as a geologic mapping tool in some areas. The following two chapters describe the distribution of radon in areas of recent deposition, a glaciated terrane and the Coastal Plain, and the controls on anomalies. Each of the last three chapters deals with a different aspect of geologic controls on radon. Included are a model for the transmission of radon from rock to water, the meteorologic controls on radon concentrations, and statistical methods to utilize indoor radon concentrations in distribution studies.

Several ground rules must be presented at the beginning of this volume. The term *radon* is used in all chapters. Radon refers to the isotope ^{222}Rn, which is a daughter product of ^{226}Ra and ultimately ^{238}U. Radon is written as Rn-222 and radon-222 in other literature. However, it is distinct from ^{220}Rn, which is referred to as thoron, because it is a daughter product of thorium.

Methods to measure indoor radon include the activated charcoal canister, alpha-sensitive track-etch film, and grab samples analyzed in an alpha scintillometer. The grab sample is a one-time measurement, the charcoal canister takes three days, and the track-etch method requires up to three months. Each has its advantages and is useful for a specific kind of analysis. Several of the chapters examine concentrations of radon in soil-gas. The technology for this analysis is relatively new.

Environmental radon researchers who study the relation between geology and radon have drawn on the techniques of the uranium exploration industry and have modified them to evaluate environmental problems better. Because uranium geologists seek to sample as large an area as possible, sampling holes are up to 2 m deep and 10 cm in diameter; gas samples are withdrawn under high vacuum pressure and circulated through an alpha scintillometer for 5 to 15 minutes. The problems with this method are that (1) the soil at the sampling site is disrupted; and (2) the sampling distance is dependent on permeability, which varies from site to site, and in many cases, is directional and controlled by fractures. This method also yields radon plus thoron because it cannot distinguish them.

A new variation on the alpha scintillometer system, developed by G. M. Reimer at the U.S. Geological Survey, has resolved many of these problems. A tensile, 3/8" hollow carbon steel probe with a sealed tip is driven 1 m into the soil using a slide hammer (Fig. 1). Approximately 20 ml of soil gas is drawn into the probe through ports near the probe tip and into a syringe through a sealed induction system attached to the top of the probe. The draw on the soil is equal to that exerted by the syringe. The sample is analyzed in an evacuated alpha scintillometer at the end of the day under static conditions. This method does not disrupt the soil and is less dependent on soil anisotropy because such a small amount of gas is withdrawn and no vacuum is used. It also yields only radon content because the thoron (half-life = 55 sec) decays long before the gas is analyzed.

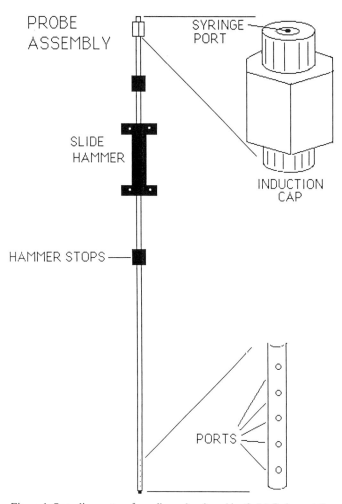

Figure 1. Sampling system for soil-gas developed by G. M. Reimer at the U.S. Geological Survey. The probe is driven into soil using the slide hammer. A syringe is inserted into the sealed induction cap and soil-gas is drawn through the ports, through the probe, and into the syringe by means of the vacuum exerted by the syringe. The syringe is sealed and the gas sample is analyzed in an alpha scintillometer later in the day.

Geology of radon in the United States

Linda C. S. Gundersen, R. Randall Schumann, James K. Otton, Russell F. Dubiel, Douglass E. Owen, and Kendell A. Dickinson
U.S. Geological Survey, Box 25046, Federal Center, MS-939, Denver, Colorado 80225

ABSTRACT

More than one-third of the United States is estimated to have high geologic radon potential. A high radon potential area is defined as an area in which the average indoor radon screening measurement is expected to be 4 pCi/L or greater. Geologic terrains of the United States with high radon potential include:

1. Uraniferous metamorphosed sediments, volcanics, and granite intrusives that are highly deformed and often sheared. Shear zones in these rocks cause the highest indoor radon problems in the United States.

2. Glacial deposits derived from uranium-bearing rocks and sediments and glacial lake deposits. Clay-rich tills and lake clays have high radon emanation because of high specific surface area and high permeability due to desiccation cracking when dry.

3. Marine black shales. The majority of black shales are moderately uraniferous and have high emanation coefficients and high fracture permeability.

4. Soils derived from carbonate, especially in karstic terrain. Although most carbonates are low in uranium, the soils derived from them are very high in uranium and radium.

5. Uraniferous fluvial, deltaic, marine, and lacustrine deposits. Much of the nation's reserve uranium ores are contained within these sedimentary deposits, which dominate the stratigraphy of the western U.S.

INTRODUCTION

Assessing the geologic radon potential of the United States has been the subject of a 2-yr project by the U.S. Geological Survey and the U.S. Environmental Protection Agency (EPA), in cooperation with the Association of American State Geologists. Indoor radon data from the State/EPA Indoor Radon Survey and other sources were compared with bedrock and surficial geology, aerial radiometric data, soil properties, and soil and water radon studies. Numeric radon and confidence indices have been developed as part of this project to quantify and standardize geologic radon potential assessment on a regional scale. Ten regional summaries on the geologic radon potential of the United States are currently in press (U.S. Geological Survey, 1992) and give detailed evaluations of the geologic radon potential of each state. The following is a summary of the methodology and findings of this study, as well as an introduction to the geology of radon.

Radon formation and migration

Radon-222 (^{222}Rn) is produced from the radioactive decay of radium-226 (^{226}Ra), which is, in turn, a product of the decay of uranium-238 (Fig. 1). Other isotopes of radon occur naturally, but are of less importance in terms of indoor radon health risk because of their short half lives and less common occurrence. A possible exception to this is thoron (^{220}Rn, a decay product of thorium), which occurs in sufficiently high concentrations to be of concern in a few local areas. In general, the concentration and the mobility of radon in a soil are dependent on several factors, the most important of which are the soil's radium content and distribution, porosity, permeability to gas movement, and moisture content. These characteristics are, in turn, determined by the character of the bedrock, glacial deposits, or transported sediments from which the soil was derived, as well as the climate and the soil's age or maturity.

Radon transport in soils occurs by two processes, (1) con-

Gundersen, L.C.S., Schumann, R. R., Otton, J. K., Dubiel, R. F., Owen, D. E., and Dickinson, K. A., 1992, Geology of radon in the United States, *in* Gates, A. E., and Gundersen, L.C.S., eds., Geologic Controls on Radon: Boulder, Colorado, Geological Society of America Special Paper 271.

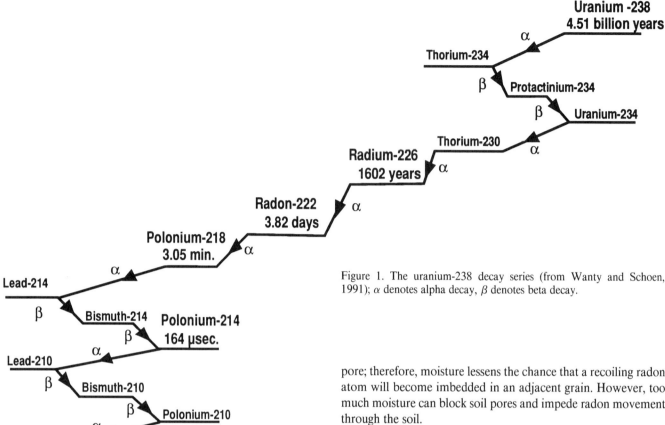

Figure 1. The uranium-238 decay series (from Wanty and Schoen, 1991); α denotes alpha decay, β denotes beta decay.

vective or advective flow, and (2) diffusion (Tanner, 1964). Diffusion is the dominant radon transport process in soils of low permeability (generally less than 10^{-7} cm^2), whereas convective transport processes tend to dominate in highly permeable soils (generally greater than 10^{-7} cm^2) (Sextro and others, 1987). Radon transport distance is limited in low-permeability soils because of the short distance radon may travel before decaying (radon has a relatively short life of 3.825 days).

When radium decays in the soil, not all of the radon produced will be mobile. The portion of radon actually released into the pores and fractures of rocks and soils is called the emanating fraction. When a radium atom decays to radon, the energy generated is strong enough to send the radon atom a distance of about 40 nm (equal to 10^{-9} m), or about 2×10^{-6} in; this is known as alpha recoil (Tanner, 1980). Depending on where radium is distributed in the soil, much of the radon produced could be imbedded in the original soil grain or in adjacent grains rather than being released to the pore space between grains. Moisture in the pore spaces greatly affects the recoil distance a radon atom will travel. Because water is more dense than air, a radon atom will travel a shorter distance in a water-filled pore than in an air-filled pore; therefore, moisture lessens the chance that a recoiling radon atom will become imbedded in an adjacent grain. However, too much moisture can block soil pores and impede radon movement through the soil.

Soil-gas radon concentrations can vary in response to climatic and weather changes on hourly, diurnal, or seasonall time scales. Schumann and others (1989a) and Rose and others (1988) recorded variations as much as an order of magnitude in soil-gas radon concentrations between seasons in Colorado and Pennsylvania, respectively. The most important factors appear to be soil moisture conditions (determined mainly by precipitation), barometric pressure, and temperature.

Radon entry into buildings

A driving force (reduced atmospheric pressure in the house relative to the soil) and entry points must exist for radon to enter a building from the soil. The negative pressure caused by furnace combustion, ventilation devices, and the stack effect during cold winter months are common driving forces. Cracks and other penetrations through building foundations, sump holes, and slab-to-foundation wall joints are common entry points. Homes with basements generally have more entry points for radon, commonly have a more pronounced stack effect, and typically have lower air pressure relative to the surrounding soil than do nonbasement homes.

Elevated levels of indoor radon occur in both basement and nonbasement homes. The term nonbasement applies to slab-on-grade or crawl space construction. In many homes with basements, the main floors have radon levels similar to those on the main floors of nonbasement homes, implying that occupants of basement homes are at greater risk from radon exposure only if they spend a significant amount of time in their basements.

Health effects from radon exposure

The health effects of radon are the subject of much debate. Recent reports of the BEIR IV and BEIR V committees (National Research Council, 1988, 1990) indicate that a portion of the population may be exposed to potentially harmful levels of radiation from radon progeny. Radon produces two short-lived polonium daughters, ^{218}Po and ^{214}Po, which area solids and tend to attach to particulates and aerosols in the air. When the polonium is inhaled, it lodges in the lung and can cause damage to the lung lining by alpha radiation as it decays. The tissues in the nasopharynx, tracheobronchial tree, and the pulmonary region receive the majority of the radiation dose, with damage to the bronchi being predominant in humans. These sites contain "precursor" or "stem" cells that are particularly sensitive to the cancer-causing properties of alpha radiation (Cross, 1987). The primary sources of epidemiologic data relating lung cancer deaths to radon exposure come from studies of underground uranium miners in the United States, Czechoslovakia, and Canada; iron, zinc, and lead miners in Sweden and Great Britain; and fluorspar miners in Newfoundland. From 3 to 8 percent of the miners studied to date developed lung cancer that is attributable to radon-progeny exposure (i.e., over and above those cancers attributed to smoking and other causes) (Cross, 1987). A number of reports summarize the available epidemiologic data from mining and residential exposure to radon progeny, including Archer and others (1976), Cross (1987), Hopke (1987), National Research Council (1988, 1990, 1991), and Steinhäusler (1988).

METHODS AND SOURCES OF DATA FOR ASSESSMENT

The assessment of geologic radon potential for the United States was made using five main types of data: (1) geologic (lithologic), (2) radiometric, (3) soil characteristics, including soil moisture and permeability, (4) indoor radon data, and (5) building architecture (specifically, whether homes in each area are predominantly slab-on-grade or crawl space construction, as opposed to homes with basements). These elements were examined and integrated to produce estimates of radon potential. This section describes these types of data and how they were used to evaluate geologic radon potential.

Geologic data

Information on the type and distribution of lithologic units and other geologic features in an assessment area is of primary importance. Rock types with naturally high uranium concentrations (considered to be greater than 2 ppm for the purposes of this assessment) that are most likely to cause indoor radon problems include carbonaceous black shales, glauconite-bearing sandstones, some fluvial sandstones, phosphorites and phosphatic sediments, chalk, some carbonates, some glacial deposits, bauxite, lignite, some coals, uranium-bearing granites and pegmatites, metamorphic rocks of granitic composition, graphitic slates, phyllites, and schists, felsic and alkalic volcaniclastic and pyroclastic volcanic rocks, syenites and carbonatites, and many sheared or faulted rocks. The most common modes of occurrence of uranium and radium within these rocks are summarized in Table 1. Rock types least likely to cause radon problems include marine quartz sands, noncarbonaceous shales and siltstones, some clays and fluvial sediments, metamorphic and igneous rocks of mafic composition, and mafic volcanic rocks. Exceptions exist within these general lithologic groups because of the occurrence of localized uranium deposits such as roll-front deposits. The most common sources of uranium and radium are the heavy minerals such as zircon, titanite, and monazite, iron-oxide coatings on rock and soil grains, and organic materials in soils and sediments. Less common are phosphate and carbonate complexes, and uranium minerals. Uranium and radium in soils are most often located on the surfaces of clays; with metal oxides, especially iron oxides; with calcium carbonate; and with organic matter.

Although many cases of extreme indoor radon levels can be traced to high radium and/or uranium concentrations in bedrock and sediments, some structural features, most notably faults and shear zones, have been identified as sites of localized uranium concentrations and have been associated with some of the highest known indoor radon levels. Two of the highest known indoor radon occurrences in the United States are associated with sheared fault zones near Boyertown, Pennsylvania (Gundersen, 1991; Gundersen and others, 1988a), and in Clinton, New Jersey (Muessig and Bell, 1988; Henry and others, 1989).

National Uranium Resource Evaluation Program (NURE) aerial radiometric data

Aerial radiometric data are used to describe the radioactivity of rocks and soils. Equivalent uranium (eU) data provide an estimate of the surficial concentrations of radon parent materials (uranium, radium) in rocks and soils. Equivalent uranium is calculated from the counts received by a gamma-ray detector in the wavelength corresponding to emissions from bismuth-214 (^{214}Bi), with the assumption that uranium and its decay products are in secular equilibrium. It is expressed in units of parts per million of uranium. Gamma radioactivity may also be expressed in terms of a radium concentration; 3 ppm eU corresponds to approximately 1 pCi/g of radium-226. ^{214}Bi is used for this purpose because it is one of the few uranium-series nuclides that emits an energetic gamma ray. Most of the ^{238}U-series radionuclides emit alpha particles or low-energy beta or gamma rays that travel only a short distance in air. Because it is the third short-lived decay product of ^{222}Rn, ^{214}Bi also provides an indication of the amount of ^{222}Rn in the near-surface soil layers.

The aerial radiometric data used for assessing radon poten-

TABLE 1. ROCKS MOST LIKELY TO CAUSE RADON PROBLEMS AND THE URANIUM AND RADIUM SOURCES THEY HOST*

Black shales, lignite, coal	Uranium-bearing organic compounds; autunite; tyuyamunite
Glauconitic sandstones	Radium and uranium-bearing iron oxides; heavy minerals
Fluvial and lacustrine sandstones	Roll-front deposits, which include uraninite, coffinite, pitchblende, secondary uranium minerals (tyuyamunite, carnotite, uranophane, and other uranyl vanadates); uranium and radium adsorbed onto organic material; iron and titanium oxides; placer deposits, which include heavy minerals
Phosphorite and phosphate	Phosphate complexes; apatite
Chalk and marl	Phosphate complexes; apatite
Carbonates	Uranium and radium adsorbed onto iron-oxide coatings; radium with organic material in soils; tyuyamunite, carnotite, and uranophane in karst and caves
Glacial deposits	Bedrock-derived clasts that comprise the glacial deposits are usually the principal source of radioactivity; uranium- and radium-bearing iron-oxide and carbonate coatings on clasts are common
Granites	Heavy minerals; uraninite; brannerite; apatite
Granitic metamorphic rocks	Heavy minerals; ultrametamorphic minerals, which include uraninite and uranothorite
Volcanic rocks	Heavy minerals; uranosilicates
Faulted rocks	Heavy minerals; uraninite; uranium precipitated with hematite and titanium oxide; minerals found in uranium vein deposits
Vein and vein-like deposits	Many kinds of uranium minerals; heavy minerals; apatite
Syenites, carbonatites, pegmatites	Uraninite; other uranium minerals; heavy minerals
Bauxite	Heavy minerals

*Compiled from Nash and others, 1981; DeVoto, 1984, 1988; Young, 1984; Smith, 1984; Nichols, 1984; Gundersen, 1989a,c.

tial in this study were collected as part of the National Uranium Resource Evaluation (NURE) program of the 1970s and early 1980s. The purpose of the NURE program was to identify and describe areas in the United States having potential uranium resources (U.S. Department of Energy, 1976). The NURE aerial radiometric data were collected by aircraft in which a gamma-ray spectrometer was mounted, flying approximately 122 m (400 ft) above the ground surface. Smoothing, filtering, recalibrating, and matching of adjacent quadrangle data sets were performed to compensate for background, altitude, calibration, and other types of errors and inconsistencies in the original data set. The corrected data were then used to produce a contour map of eU values for the conterminous United States (Duval and others, 1989a).

Although radon is highly mobile in soil, and its concentration is affected by meteorologic conditions (Tanner, 1980; Kovach, 1945; Klusman and Jaacks, 1987; Schery and others, 1984), relatively good correlations between average soil-gas radon concentrations and average eU values for some soils have been noted (Gundersen and others, 1988a, b; Schumann and Owen, 1988). The shallow (20 to 25 cm) depth of investigation of gamma-ray spectrometers, either ground-based or airborne (Duval and others, 1971; Durrance, 1986), suggests that gamma-ray data may sometimes provide an underestimate of radon source strength in soils in which some of the radionuclides in the near-surface soil layers have been transported downward through the soil profile or depleted by other processes. The redistribution of radionuclides in soil profiles is dependent on a combination of climatic, geologic, and geochemical factors. Given sufficient understanding of the factors involved, these regional differences may be predictable.

Soil survey data

Soil surveys prepared by the U.S. Soil Conservation Service (SCS) provide data on soil characteristics. The reports are commonly available in county formats and state summaries, and usually contain both generalized and relatively detailed maps of soils in the area. Because of time and map-scale constraints, it was impractical to examine county soil reports for each county in the United States, so more generalized summaries at appropriate scales were used where available. For state or regional-scale radon characterizations, soil maps are compared to geologic maps of the area, and the soil descriptions, shrink-swell potential, depth to seasonal high water table, permeability, and other relevant characteristics of each soil group noted. One of the best summaries of the national distribution of technical soil types is the Soils Sheet of the U.S. Geological Survey National Atlas of the United States (U.S. Soil Conservation Service, 1987).

As soils form, they develop distinct layers, or horizons, that are cumulatively called the soil profile. The A horizon is a surface horizon containing a relative abundance of organic matter but dominated by mineral matter. The B horizon underlies the A horizon. Important characteristics of B horizons include accumulation of clays, iron oxides, calcium carbonate or soluble salts, and organic matter complexes. In drier climates, a horizon may exist within or below the B horizon that is dominated by calcium carbonate, often called caliche or calcrete. This carbonate-cemented horizon is designated the K horizon. The C horizon underlies the B and is a zone of weathered parent material. It is generally not a zone of significant leaching or accumulation.

Soil permeability is typically expressed in SCS soil surveys in terms of the speed, in inches per hour, at which water soaks into the soil, as measured in a soil percolation test. Although inches per hour is not truly a unit of permeability, it is in widespread use and is referred to as "permeability" in SCS soil surveys. The permeabilities listed in the SCS surveys are for water, but they can generally be applied to gas permeability in unsaturated soils. Permeabilities greater than 6.0 in/hr may be considered high, and permeabilities less than 0.6 in/hr may be considered low in terms of soil-gas transport.

Many well-developed soils contain a clay-rich B horizon that may impede vertical soil gas transport. Radon generated below this horizon cannot readily escape to the surface, so it would instead tend to move laterally, especially under the influence of a negative pressure exerted by a building. Depth to seasonal high water table can also be an important parameter to consider in some areas. Because water in soil pores inhibits gas transport, the amount of radon available to a home is effectively reduced by a high water table. Areas likely to have high water tables are river valleys, coastal areas, and some areas overlain by deposits of glacial origin (e.g., loess).

Shrink-swell potential is an indicator of the abundance of smectitic (swelling) clays in a soil. Soils with a high shrink-swell potential may cause building foundations to crack, and thus create pathways for radon entry into the structure. In addition, swelling soils often crack as they dry; as a result, they provide additional pathways for soil-gas transport and effectively increase the gas permeability of the soil (Schumann and others, 1989a, 1990).

Indoor radon data

Three major sources of indoor radon data were used. All data reflect screening measurements collected with charcoal cannister detectors. The first and largest source of data is from the EPA/State Residential Radon Surveys (Ronca-Battista and others, 1988; Dziuban and others, 1990). During the period 1986-1992, 40 states completed EPA-sponsored radon surveys. The EPA/State Residential Radon Surveys were designed to be comprehensive and statistically significant at the state level, and were subjected to high levels of quality assurance and control. The surveys collected 2- to 7-day screening measurements in the lowest liveable areas of the home. The target population for the surveys included owner-occupied, single family, detached housing units (White and others, 1989). Participants were selected randomly from telephone directory listings. The surveys have yielded approximately 57,000 data points through 1991.

The second source of indoor radon data comes from residential surveys that have been conducted either within a specific state or region of the country (e.g., independent state surveys or utility company surveys). Examples include indoor radon surveys conducted by the states of Florida, New York, New Jersey, and Utah.

The third source of indoor radon data comes from the University of Pittsburgh Radon Project (Cohen and Gromicko, 1988; Cohen, 1990). This ongoing effort has been collecting indoor radon data since 1985, and to date has accumulated approximately 175,000 indoor radon measurements for the entire country. The data are collected from homeowners who purchase radon detectors from the project on a voluntary basis. The data are therefore nonrandom, although a limited bias reduction technique has been applied to the data set by eliminating certain values based on responses to a standard questionnaire (Cohen, 1990).

Table 2 is a summary of the EPA/State Residential Radon Survey data. The arithmetic means for the states range from a high of 8.8 pCi/L in Iowa to a low of 0.1 pCi/L in Hawaii. In 15 states, 20 percent or more of those homes measured were above 4 pCi/L, whereas 10 states had 10 percent or fewer of the homes measured above the action guideline. Although a state survey

TABLE 2. INDOOR RADON DATA FROM STATE/EPA RESIDENTIAL RADON SURVEY

State	Average	% > 4 pCi/L	Rank
Alaska	1.7	7.7	25
Alabama	1.8	6.4	29
Arizona	1.6	6.5	28
California	0.9	2.4	32
Colorado	5.2	41.5	5
Connecticut	2.9	18.5	17
Georgia	1.8	7.5	26
Hawaii	0.1	0.4	34
Iowa	8.8	71.1	1
Idaho	3.5	19.3	16
Indiana	3.7	28.5	9
Kansas	3.1	22.5	13
Kentucky	2.7	17.1	18
Louisiana	0.5	0.8	33
Maine	4.1	29.9	7
Massachusetts	3.4	22.7	12
Michigan	2.1	11.7	23
Minnesota	4.8	45.4	4
Missouri	2.6	17.0	19
North Carolina	1.4	6.7	27
North Dakota	7.0	60.7	2
Nebraska	5.5	53.5	3
New Mexico	3.1	21.8	14
Nevada	2.0	10.2	24
Ohio	4.3	29.0	8
Oklahoma	1.1	3.3	31
Pennsylvania	7.7	40.5	6
Rhode Island	3.2	20.6	15
South Carolina	1.1	3.7	30
Tennessee	2.7	15.8	21
Vermont	2.5	15.9	20
Wisconsin	3.4	26.6	10
West Virginia	2.6	15.7	22
Wyoming	3.6	26.2	11

may reflect a low percentage of homes with elevated levels of radon, this does not necessarily suggest that an insignificant number of homes could be affected by elevated radon levels. In California, for example, only 2.4 percent of the homes measured had indoor radon levels greater than 4 pCi/L. This small percentage is misleading, however, because it represents an overall total of more than 247,000 homes that may be effected by elevated levels.

Radon index and confidence index

Many of the geologic methods used to evaluate an area for radon potential require subjective opinions based on the professional judgment and experience of the individual geologist. However, these evaluations are based on established scientific principles that are universal to any geographic area or geologic setting. This section describes the methods and conceptual framework used to evaluate areas for radon potential based on the geologic factors discussed in the previous sections. The scheme is divided into two basic parts, a Radon Index, used to rank the general radon potential of the area, and the Confidence Index, used to express the level of confidence in the prediction based on the quantity and quality of the data used to make the determination. This scheme works best if the areas to be evaluated are delineated by geologically based boundaries (geologic provinces) rather than political ones (state/county boundaries) in which the geology may be varied across the area.

Radon index. Table 3 presents the Radon Index (RI) matrix. Five main categories are evaluated and a point value of 1, 2, or 3 is assigned to each category. These categories were selected because the factors they represent are considered to be of primary importance in controlling radon potential and because at least some data for these factors are consistently available for every geologic province. Because each of these main factors encompasses a wide variety of complex and variable components, the subjective professional judgment and experience of the geologists performing the evaluation are heavily relied on in assigning point values to each category.

Indoor radon is evaluated using unweighted arithmetic means of the indoor radon data for each county or for each geologic area to be evaluated. Other expressions of indoor radon levels in an area could also be used, such as weighted averages, living area averages, or annual averages.

Aerial radioactivity data used in this report are from the equivalent uranium map of the conterminous United States compiled from NURE aerial gamma-ray surveys (Duval and others, 1989a). An average eU value was estimated visually for each selected geologic area and applied to the matrix to determine its factor score.

The geology factor is complex and actually incorporates many geologic characteristics. In the matrix, the terms positive and negative refer to the presence or absence and distribution of rock types known to have high uranium contents and to generate elevated radon in soils or indoors. Examples of positive rock types include granites, black shales, and phosphatic rocks. Examples of negative rock types include quartz sands and some clays. The term variable indicates that the geology within the region is variable or that the rock types in the area are known or suspected to generate elevated radon in some areas, but not in others, due to compositional differences, climatic effects, localized distribution of uranium, or other factors. Geologic information indicates not only how much uranium is present in the rocks and soils, but also gives clues for predicting general radon emanation and mobility characteristics through additional factors such as rock and soil geochemical characteristics and structure (notably the presence of faults or shears).

To add additional weight to the geologic factor in cases where additional reinforcing or contradictory geologic evidence is available, Geologic Field Evidence (GFE) points are added or subtracted from an area's score. Relevant geologic field studies are important in enhancing our understanding of how geologic processes affect radon distribution. In some cases, geologic models and supporting field data reinforce an already strong (high or low) score; in others, they can provide important contradictory data. For example, areas of the Dakotas, Minnesota, and Iowa that are covered with Wisconsin-age glacial deposits exhibit a low aerial radiometric signature and would score only one point in that category. However, data from geologic field studies in North Dakota and Minnesota (Schumann and others, 1990) suggest that eU is a poor predictor of geologic radon potential in this area because radionuclides have been leached from the upper soil layers, but are present and possibly even concentrated in deeper soil horizons, generating significant soil-gas radon. This positive supporting field evidence adds two GFE points to the score, which helps to counteract the invalid conclusion suggested by the radiometric data. No GFE points are awarded if there are no documented field studies for the area.

The term soil permeability refers to several soil characteristics that influence radon concentration and mobility, including soil type, grain size, structure, soil moisture, drainage, and permeability. In the matrix, the term low refers to permeabilities less than about 0.6 in/hr; the term high corresponds to greater than about 6.0 in/hr, in SCS standard soil percolation tests.

Architecture type refers to whether homes in the area have mostly basements, mostly slab-on-grade construction, or a mixture of the two. Split-level and crawl space homes fall into a category termed mixed.

Confidence index. Except for architecture type, the same factors are used to establish a Confidence Index (CI) for the radon potential prediction for each area (Table 4). Architecture type is not included in the confidence index because house construction data are readily and reliably available through surveys taken by agencies and industry groups including the Bureau of the Census, National Association of Home Builders, and the Federal Housing Administration. The remaining factors are scored on the basis of the quality and quantity of data used in the RI matrix.

Indoor radon data are evaluated on the distribution and number of data points and on whether the data are consistent,

TABLE 3. RADON INDEX MATRIX*

Point value	Indoor radon (avg.)	Aerial radioactivity	Geology†	Soil permeability	Architecture type
1	< 2 pCi/L	< 1.5 ppm eU	negative	low	mostly slab
2	2 – 4 pCi/L	1.5 – 2.5 ppm	variable	moderate	mixed
3	> 4 pCi/L	> 2.5 ppm	positive	high	mostly basement

*Scoring key: low Rn = 3–8 points (< 2 pCi/L probable indoor avg.); moderate Rn = 9–11 points (2–4 pCi/L probabale indoor average); high = 12–17 points (> 4 pCi/L probable indoor average); total possible range of points = 3 to 17.

†Geologic field evidence (GFE) points: GFE points are assigned in addition to points for the "Geology" factor for specific, relevant geologic field studeis; see text for details.
Geologic evidence supporting: high radon = 2 points; moderate radon = +1 point; low radon = -2 points; no relevant geologic field studies = 0 points.

TABLE 4. CONFIDENCE INDEX MATRIX*

Point value	Indoor radon data	Aerial radioactivity	Geologic data	Soil permeability
1	Sparse / no data	Questionable / no data	Questionable	Questionable / no data
2	Fair coverage / quality	Glacial cover	Variable	Variable
3	Good coverage / quality	No glacial cover	Proven geologic model	Reliable, abundant

*Scoring key: low confidence = 4 to 6 points; moderate confidence = 7 to 9 points; high confidence = 10 to 12 points. Total possible range of points = 4 to 12.

randomly sampled data (State/EPA Indoor Radon Survey or other State survey data) or volunteered vendor data (likely to be nonrandom and biased toward population centers and/or high indoor radon levels). The categories listed in the CI matrix reflect the levels of sampling density and statistical robustness of an indoor radon data set.

Aerial radioactivity data are available for all but a few areas of the continental United States and for part of Alaska. In general, the greatest problems with correlations among eU, geology, and soil-gas or indoor radon levels appear to be associated with glacial deposits. Correlations among eU, geology, and radon are generally sound in unglaciated areas; these areas are usually assigned three CI points. Radioactivity data in some unglaciated areas may be assigned fewer than three points, and in glaciated areas assigned only one point, if the data are considered questionable or coverage is poor.

For the geologic data factor, a high confidence score is given to an area where a proven geologic model for radon generation and mobility can be applied. Rocks for which the processes are less well known or for which data are contradictory are regarded as variable; those about which little is known or for which no apparent correlations have been found are deemed questionable.

The soil permeability factor is also scored on quality and amount of data. Soil permeability can be approximated from grain size and drainage class if data from standard, accepted soil percolation tests are unavailable. Percolation test data and other measured permeability data are more accurate and score a higher confidence level.

Examples of radon potential ratings applied to different geologic terrains in the United States are given in Table 5. Space does not permit including the rating tables for all of the United States.

RADON POTENTIAL IN THE UNITED STATES

Areas of the United States that are geologically similar can be grouped and delineated on a map (Fig. 2). Each region is characterized by a basic geology and climate that determines its radon potential. By examining and correlating available geologic, aerial radiometric, soil radon, and indoor radon data, generalized estimates of the radon potential of each region can be made. The following is a discussion of major geologic features and rock types and their known or expected radon potential for each geologic/physiographic region. In each case, large-scale, well-known, or highly anomalous features are discussed. This list is by no means exhaustive; rather, it is intended to give the reader a general feeling for the geologic features in each area that are likely to produce elevated indoor radon values, point out important rock units or other geologic features where they are known, and act as a general guide for using geology to predict radon potential on a regional scale. The numbered regions designated in the text correspond with those in Figure 2.

Regions 1, 2, 3: Outer and Inner Coastal Plain and phosphatic and limestone deposits of Florida

The Coastal Plain of the eastern and southern United States consists of a systematic progression of predominantly marine and

TABLE 5. RADON INDEX AND CONFIDENCE INDEX EXAMPLES

Factor	Des Moines Lobe region		Superior Upland		Michigan Basin		Great Plains		Northern Coast Ranges and Cascade Mtns.		Puget Lowlands		Inner Coastal Plain		Appalachian Carbonates		Basin and Range	
	RI	CI	RI	CI	RI	CI	RI	CI	RI	CI	RI	CI	RI	CI	RI	CI	RI	CI
Indoor radon	3	3	2	3	2	3	2	3	1	2	1	2	2	3	3	3	2	2
Radioactivity	1	2	1	2	1	2	2	3	1	2	1	3	2	3	2	2	2	3
Geology	3	3	2	2	2	3	2	2	1	2	1	2	2	3	3	3	2	2
Soil permeability	1	2	2	2	2	2	2	2	2	2	2	3	2	3	2	2	2	2
Architecture	3	1	3	1	3	1	3	1	1	1	1	1	1	1	3	1	2	1
GFE points	+2	1	0	1	0	1	0	1	0	1	0	1	1	1	3	1	0	1
Total	13	10	10	9	10	10	11	10	6	8	6	10	10	12	16	10	10	9
	High	High	Mod	Mod	Mod	High	Mod	High	Low	Mod	Low	High	Mod	High	High	High	Mod	Mod

fluvial sediments deposited during the evolution of the Atlantic and Gulf Coasts. The oldest rocks exposed in the Coastal Plain are Cretaceous in age and consist predominantly of glauconitic sandstones, chalks, and clays, as well as some nonglauconitic quartz sandstones and fossiliferous limestone. These are overlain by lower Tertiary (Paleocene, Eocene, and Oligocene) sands and clays, which are often glauconitic, and upper Tertiary (Miocene) fossiliferous chalks, clays, and thin sands. The youngest Tertiary sediments (Pliocene) are dominated by gravelly sands, clayey sands, and thin clay beds. Because of the consistency in the general stratigraphy of the Coastal Plain, many of the lithologic sequences are similar from state to state. Soil radon, surface radioactivity, uranium and radium concentrations, permeability, and soil grain size distributions have been measured along more than 1,600 km of transects in five states underlain by Coastal Plain sediments (Peake and others, 1988; Peake and Gundersen, 1989). In general, the data suggest that the Inner Coastal Plain (Region 2), which is composed of Cretaceous and Early Tertiary sediments, has higher radon potential than the Outer Coastal Plain (Region 1), which is composed of Middle to Late Tertiary and Quaternary sediments. Grab samples of radon in soil gas collected at a depth of 1 m averaged 700 to 1,000 pCi/L for the Coastal Plain as a whole. The two highest soil radon measurements were taken in Inner Coastal Plain sediments; 16,226 pCi/L was measured in the glauconitic sands of the Nevasink Formation in New Jersey and 6,333 pCi/l was measured in the carbonaceous shales of the Eagle Ford Group in Texas. Radon in soil gas greater than 1,000 pCi/L associated with phosphatic fossil layers and glauconitic sands and clays in the Aquia, Brightseat, and Calvert Formations in Maryland and Virginia have also been reported (Otton, 1991). Localized concentrations of uranium, found in roll-front uranium deposits in Texas and in marine sands and heavy mineral deposits from Virginia to Georgia, have produced some locally high indoor radon occurrences. Heavy mineral deposits found throughout the Coastal Plain also have the potential for creating scattered local anomalies and are a potential source of thoron as well.

Comparisons of indoor radon data from the State/EPA Indoor Radon Survey (winter screening measurements from 1986–1989) with other data sources show good correlations among soil radon, radionuclide data, and indoor radon data. The average for indoor radon concentrations is 1 pCi/L or less over different parts of the outer Coastal Plain. Areas underlain by Cretaceous chalks, carbonaceous shales, phosphatic sediments, and glauconitic sandstones of the Inner Coastal Plain average 2.3 pCi/L and have the highest radon potential.

Region 3 in Florida outlines the general extent of the uraniferous phosphatic deposits that cause abundant, moderate radon problems and locally high radon problems. The geologic units thought responsible for these problems include the Hawthorn Formation (Smith and Hansen, 1989) and the Alachua and Bone Valley Formations. The southern part of Region 3 in Dade County, Florida, may be moderate in radon potential due to the Key Largo Limestone.

Regions 4, 5, 6, and 7: Northern Appalachian Mountains including New England; Taconic, Adirondack, and Green Mountains; and Northern Appalachian Plateau

Region 4 is made up of Proterozoic and Paleozoic metamorphic and igneous rocks of moderate radon potential, the main source of radon being uraniferous minerals and faults. Volcanic units in this region are low in radon potential, while the schists, gneisses, and granites are dominantly felsic in composition and produce moderate radon concentrations. Glacial tills and gravels have compositions derived from both local and northern rock sources and create locally high radon due to permeability and a moderate uranium source.

Region 5 is also an area of crystalline bedrock with locally derived glacial tills and gravels. The granites of this area, particularly the Conway Granite, have very high uranium concentrations. Not only do the granites, associated Proterozoic metamorphic rocks, and fault zones cause high indoor radon, they also cause extremely high radon in ground water, with domestic well water concentrations in the range of 1 million pCi/L (Hall and others, 1987) occurring in the fractured granite aquifers.

Region 6 is a highly variable terrain that includes the metamorphic and igneous rocks of the Green Mountains and the Paleozoic rocks of the northern Appalachian Mountains, Appalachian Plateau, and Taconic Mountains. The Paleozoic rocks of the Appalachians include highly deformed shales, sandstones, and carbonates. Some of the carbonates and shales cause locally moderate to high indoor radon. Gravels and tills in the Albany area, and graphitic schists, phyllite, and slate in the Taconic and Green Mountains also cause locally high indoor radon.

Region 7, the Adirondacks, is a region of metamorphic and igneous rocks of contrasting radon potential. The Marcy Anorthosite complex, which forms the core of the Adirondack Mountains, is low in radon potential, whereas the metamorphic schists and gneisses that blanket the rim of the Adirondacks are locally high in radon potential due to uraniferous minerals, uranium deposits associated with several of the magnetite deposits, and shear zones. Gravels and tills cause locally high radon.

Region 8: Central and southern Appalachian Mountains, including Mesozoic Basins, Piedmont, Blue Ridge, and Valley and Ridge

The eastern part of the Appalachian mountains, known as the Piedmont and Blue Ridge, is underlain by Proterozoic and Paleozoic-age metamorphic and igneous rocks. These rocks have

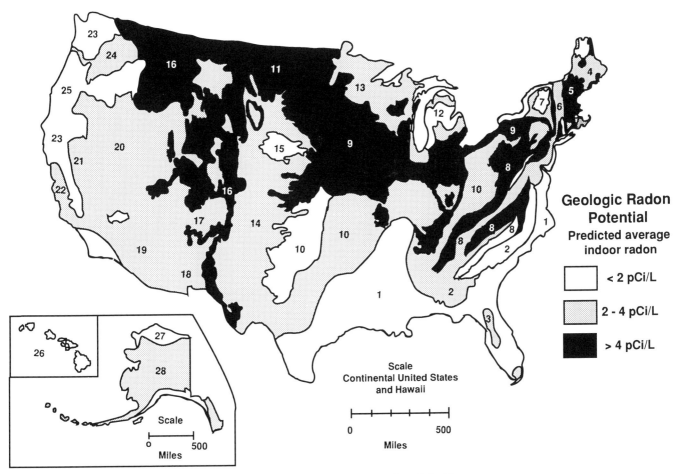

Figure 2. Geologic radon potential map of the United States.

moderate radon potential, with localized areas of high potential, especially in faults and granitic rocks. Studies thus far have yielded an average of 1,000 pCi/L for rocks of granitic composition and an average of 600 pCi/L for rocks of mafic composition. More than a thousand indoor and soil-gas radon measurements have been averaged for geologic units in the Appalachian region of Pennsylvania, New Jersey, Maryland, and Virginia, and indicate that, on the average, the indoor radon concentration is approximately one percent of the soil radon concentration (Gundersen, 1989a). Permeability and emanating power are the main factors affecting this relationship.

Early Mesozoic sedimentary rocks underlie the eastern part of this region from Massachusetts to northern South Carolina (Froelich and Robinson, 1988), and have variably low to high radon potential. Anomalous uranium occurs in fluvial sandstones and conglomerates (Turner-Peterson, 1988), including the New Haven Arkose (Hartford Basin), Stockton Formation (Newark Basin), New Oxford Formation (Gettysburg Basin), Manassas Sandstone (Culpepper Basin), and part of the Cow Branch Formation (Dan River–Danville Basin). Diabase intrusions are common within Triassic sedimentary rocks in some of the Early Mesozoic basins. Although the diabase is generally low in radon potential, the hornfels formed by contact metamorphism of the surrounding siltstones and sandstones have moderate to high radon potential (Gundersen and others, 1988b). Uraniferous black shales in the early Mesozoic basins also have locally high radon potential (Szabo and Zapecza, 1991). In contrast to the marine black shales in most of the United States, uranium-bearing black shales in the early Mesozoic basins are of lacustrine origin. Uraniferous lacustrine black shales occur in the Lockatong and Passaic Formations (Newark Basin), Gettysburg Formation (Gettysburg Basin), and the Balls Bluff Siltstone (Culpepper Basin). The Cumnock Formation in the Deep River Basin contains uranium-bearing phosphatic layers within the black shales.

Paleozoic-age sedimentary rocks cover an extensive area of the western Appalachians, known as the Valley and Ridge, and consist of sandstones, siltstones, shales, and carbonates. The carbonate soils, black shale soils, and black shale bedrock can generate moderate to high levels of indoor radon. Carbonate soils derived from Cambrian-Ordovician rock units of the Valley and Ridge Province cause known indoor radon problems in eastern Tennessee, western New Jersey, western Virginia, eastern West Virginia (Schultz and Brower, 1991; Schultz and Brower, this volume), and central and eastern Pennsylvania. The carbonate rocks themselves are low in uranium and radium; however, the soils developed on these rocks are derived from the dissolution of the $CaCO_3$ that makes up the majority of the rock. When the $CaCO_3$ has been dissolved away, the soils are enriched in the remaining impurities, predominantly base metals, including uranium. Rinds containing high concentrations of uranium and uranium minerals can be formed on the surfaces of rocks involved with $CaCO_3$ dissolution. Ground water derived from these areas, however, often contains radon concentrations of 1,000 pCi/L or less (Schultz and Wiggs, 1989; Smith, 1984). Carbonates also form karst topography, characterized by solution cavities, sinkholes, and caves, which increase the overall permeability of the rocks in these areas and may induce convective flow of radon.

In the Appalachians, the highest indoor, soil, and water radon values are most often associated with faults and fractures in the rock (Smith, 1984; Hall and others, 1987; Gundersen, 1989b; Gates and Gundersen, 1989).

Region 10: Nonglaciated portion of Appalachian Plateau

The Appalachian Plateau Region contains areas of moderate, and some locally high, radon potential. The carbonate soils and shales found in the Paleozoic-age domes and basins characteristic of this part of the United States have moderate to high radon potential. Of specific interest are the uranium-bearing Chattanooga shale in Kentucky (included in Region 9, although technically it lies in the Appalachian Plateau region) and Tennessee (Reesman, 1988), the Devonian-Mississippian black shales in Ohio, Pennsylvania, New York, and Indiana, and the Ordovician, Mississippian, Permian, and Pennsylvanian-age carbonates and black shales in Alabama, Indiana, Tennessee, Kentucky, Michigan, Illinois, Missouri, Iowa, Arkansas, and Oklahoma (Coveney and others, 1988). Although exposed in a limited area, Precambrian granites of the St. Francois Mountains in southeastern Missouri are among the most highly uraniferous igneous rocks in the United States (Kisvarsanyi, 1988). Granites, rhyolites and related dike rocks in the Wichita Mountains of Oklahoma have also been evaluated as having moderate to high radon potential (Flood and others, 1990). A large area of low radon potential is underlain by the Pennsylvanian Pottsville Sandstone, which extends from eastern Ohio through West Virginia, eastern Kentucky, east-central Tennessee, and northern Alabama. Moderate indoor radon is also associated with uraniferous coal deposits in Pennsylvania and West Virginia.

Regions 9, 11, 12, 13: Northern Great Plains and Great Lakes

The Northern Great Plains–Great Lakes region is underlain by Wisconsinan and pre-Wisconsinan–age glacial deposits and loess. Much of North and South Dakota, western and southern Minnesota, and northern Iowa (Region 11) are underlain by deposits of the Des Moines lobe. Des Moines lobe tills are silty clays and clays derived from the Pierre Shale and from Tertiary sandstones and shales that have relatively high concentrations of uranium and high radon emanating power. Included within this region are clay and silt deposits of glacial lakes Agassiz, Dakota, Souris, and Devil's Lake, which generate some of the highest radon levels in the area. Southwestern North Dakota is underlain by unglaciated Tertiary sandstones, siltstones, and shales, some of which include uraniferous coals and carbonaceous shales (Region 14).

In Region 9, glacial deposits in southern Wisconsin, northern Illinois, and western Indiana are primarily from the Green

Bay and Michigan lobes. These tills range from sandy to clayey and are derived mostly from sandstones and carbonate rocks of southern Wisconsin and the Illinois Basin. Eastern Indiana and western Ohio are underlain by tills derived from the Ohio and New Albany black shales. Black shales extend south of the glacial limit, forming an arcuate pattern in northern Kentucky. They also underlie and provide source material for glacial deposits in a roughly north-south pattern through central Ohio, including the Columbus area, and extend eastward into southern New York. The overall radon potential of this area is high.

In Region 12, the Michigan Basin area includes silty and clayey tills in northern Michigan and surrounding Lake Michigan. Source rocks for these tills are sandstones, shales, and carbonate rocks of the Michigan Basin, which are generally poor radon sources. Exposed crystalline rocks in the central part of the Upper Peninsula of Michigan cause locally high indoor radon levels. This area has a low to moderate overall radon potential.

The Superior Upland of Region 13 includes glacial deposits of the Lake Superior lobe in northern Minnesota and northern Wisconsin. The underlying source rocks for these tills are volcanic rocks and mafic metamorphic and granitic rocks of the Canadian Shield that have relatively low uranium contents. The sandy tills derived from these rocks have relatively high permeability, but because of their lower uranium content and lower emanating power, they have a moderate radon potential. In central Wisconsin, uraniferous granites of the Wolf River and Wausau plutons are exposed at the surface or covered by a thin layer of glacial deposits and cause some of the highest indoor radon concentrations in the state.

Glaciated areas present special problems for assessment because bedrock material may be transported hundreds of kilometers from its source and deposited as till. Glaciers are quite effective in redistributing uranium-rich rocks; for example, in Ohio, uranium-bearing black shales have been disseminated over much of the western part of the state, now covering a much larger area than their original outcrop pattern, and display a prominent radiometric high on the radioactivity map of the United States. The physical, chemical, and drainage characteristics of soils formed from glacial deposits vary according to source bedrock type and the glacial features on which they are formed. For example, soils formed from outwash or ground moraine deposits tend to be more poorly drained and contain more fine-grained material than soils formed on lateral and terminal moraines or eskers, which are generally coarser and well drained. In general, soils developed from glacial deposits are poorly structured, poorly sorted, and poorly developed, but are generally moderately to highly permeable and are rapidly weathered, because the action of physical crushing and grinding of the rocks to form tills may enhance and speed up soil weathering processes (Jenny, 1935). Clayey tills, such as those underlying most of North Dakota and a large part of Minnesota, have high emanation coefficients (Grasty, 1989) and usually have low to moderate permeability because they are mixed with coarser sediments. Tills consisting of mostly coarse material tend to emanate less radon because larger grains have lower surface area-to-volume ratios, but because these soils have generally high permeabilities, radon transport distances are generally longer, and structures built in these materials are able to draw soil air from a larger source volume, so moderately elevated indoor radon concentrations may be achieved from soils of comparatively lower radioactivity (Kunz and others, 1989; Schumann and others, 1990).

Regions 14, 15: Unglaciated Great Plains

The Great Plains extend from eastern Montana to central Texas. The area is mostly underlain by the Cretaceous Pierre Shale and by Tertiary continental sandstones, siltstones, and shales including the White River, Ogallala, and Arikaree Formations. The lower part of the Pierre Shale has an overall higher uranium content than the upper part, and locally contains black shales. Members of the White River Formation are significant radon producers in the northern and central Great Plains, whereas the Ogallala and Arikaree Formations are principal sources for indoor radon in the central and southern part of the region from Colorado to west Texas. Carbonaceous shales and uranium-bearing coals in the Tertiary sedimentary rocks of unglaciated southwestern North Dakota generate locally very high radon levels. Also included in this area are the Black Hills of southwestern South Dakota, which are underlain by Precambrian granitic and metamorphic rocks and Paleozoic sedimentary rocks of moderate radon potential. Overall, the Great Plains has a moderate radon potential.

The Sand Hills in northern Nebraska (Region 15), consisting primarily of windblown quartz sands, have a low radon potential.

Region 16: Rocky Mountains and parts of western Great Plains

The Rocky Mountains have a high radon potential similar to the Appalachian region for many of the same reasons. The metamorphic and igneous rocks in the Rocky Mountains are generally similar in composition, degree of deformation, and intrusion by granites to those of the Appalachians. However, the Rocky Mountains have undergone several periods of intense and widespread hydrothermal activity, creating vein deposits of uranium that cause local, high concentrations of indoor radon and radon in water in Colorado and Idaho (Schumann and others, 1989b; Lawrence and others, 1989). Colluvium and alluvium derived from crystalline rocks of the Rocky Mountains cover much of the plains along the Front Range from New Mexico to Canada and cause known moderate to high indoor radon problems in Colorado and Idaho (Ogden and others, 1987; Otton and others, 1988). The Northern Rocky Mountains comprise the northeast and north-central part of Washington and northern and central Idaho. This area is underlain by Precambrian sedimentary rocks, Paleozoic sedimentary rocks, and Mesozoic metamorphic rocks, all intruded by Mesozoic and Tertiary granitic rocks. The largest intrusive body, the Idaho Batholith, is a complex of granitic rock units ranging from diorite to granite. Uraniferous late Cretaceous to early Tertiary granites occur throughout the Northern Rocky Mountains. An extensive, though dissected, veneer of Tertiary volcanic rocks underlies much of the central Idaho portion of the Northern Rocky Mountains. Included in Region 16 are the Permian marine limestones and other marine sedimentary rocks of eastern New Mexico and the west Texas Panhandle. These units, in particular the Cutler Formation, Sangre de Cristo Formation, and San Andres Limestone, have the potential to create moderate and locally high indoor radon problems (McLemore and Hawley, 1988). Also included in Region 16 is an apron of Tertiary and Cretaceous sedimentary rocks with high radon potential that contain local uranium deposits and are overlain partly by colluvium from the Proterozoic rocks of the Rocky Mountains.

Region 17: Colorado Plateau, Wyoming Basin

The Colorado Plateau and Wyoming Basin are underlain by sedimentary rocks ranging in age from Pennsylvanian to Tertiary. The majority of the sedimentary uranium deposits of the United States are located in this region and high indoor radon in Utah, Colorado, Wyoming, and New Mexico appear to correspond to the uranium deposits. Dominant rock types include arkosic conglomerates, marine limestones and shales, marginal marine sandstones and shales, and fluvial and lacustrine sandstones, shales, and limestones. Uranium occurs to some extent in most of these rock types. The most significant uranium deposits occur in Mesozoic sedimentary rocks, with the Jurassic sandstones being the most common host for uranium ore. Localized sandstone-type uranium deposits are hosted, in particular, by the Triassic Chinle, the Jurassic Morrison, and the Cretaceous Dakota Formations in this region. Mine tailings from these sedimentary deposits caused some of the earliest detected indoor radon problems (Spitz and others, 1980). A small part of the Colorado Plateau is included in Region 18 and is underlain by sedimentary rocks with moderate rather than high radon potential.

In the Wyoming Basin, the Permian Phosphoria Formation has moderate to high radon potential. It covers an area of 350,000 km^2 in southeastern Idaho, northeastern Utah, western Wyoming, and southwestern Montana, and has a uranium content that varies from 0.001 to 0.65 percent. Other rocks with high radon potential in the Wyoming Basin are the Cretaceous Mancos Shale, which is uraniferous in places, and Tertiary sandstones, siltstones, and shale, which host uranium deposits and uranium-bearing coals.

Regions 18, 19, 20: Basin and Range

The Basin and Range (regions 18, 19, 20) is composed of Precambrian metamorphic rocks, late Precambrian and Paleozoic metamorphosed and unmetamorphosed sedimentary rocks, Mesozoic and Tertiary intrusive rocks, and Tertiary sedimentary and volcanic rocks. The region is structurally complex, with the aforementioned rocks forming the mountain ranges and alluvium derived from the ranges filling the basins. The sedimentary rocks include marine carbonates and shales, cherts, quartzites, and sandstones, as well as fluvial and continental sandstones, siltstones, and shales. As with the Colorado Plateau, local uranium deposits occur throughout the sedimentary rocks. Areas with moderate and locally high radon potential include the Tertiary volcanic rocks, particularly the Miocene- and Pliocene rocks that are found throughout the Basin and Range Province, Precambrian gneiss in southern Nevada, and the Carson Valley alluvium, which is derived from the Sierra Nevada uraniferous granites. In Utah, Sprinkel (1988) has indicated that the Wasatch fault zone and some geothermal areas have the potential to produce elevated radon. The southern part of the Basin and Range (Region 19) has fewer Tertiary volcanic rocks and is notably lower in radon potential.

The Snake River Plain in the northern part of Region 20 forms an arcuate depression in southern Idaho underlain by basaltic volcanic rocks. Alluvium from adjacent mountains and tuffaceous sedimentary rocks underlies much of the upper Snake River Valley and the western end of the Snake River Plain. Those areas underlain by basalt have low to locally moderate radon potential; however, those areas underlain by tuffaceous sedimentary rocks and alluvium along the Snake River Valley have high overall radon potential. Overall, the area has a moderate radon potential.

Regions 21, 22: Sierra Nevada, Great Valley, and Southern Coast Ranges

The Sierra Nevada (Region 21) is underlain by Paleozoic and Mesozoic metamorphic rocks, with the metamorphic rocks dominant in the northern part of the range and the granites dominant in the southern part of the range. Tertiary volcanic rocks are also found in the northern part of the range. The granites of the Sierra Nevada Mountains are high in uranium and have moderate radon potential, as does the colluvium formed on the eastern and western flanks of the mountains. The granite and colluvium are associated with high indoor radon in Nevada as well as California.

The Southern Coast Ranges include the Franciscan Formation, a complex assemblage of metamorphosed marine sedimentary rocks and ultramafic rocks, Cretaceous and Tertiary sedimentary rocks, and Mesozoic metamorphic and igneous rocks. The Tertiary marine sedimentary rocks and Mesozoic igneous and metamorphic rocks are uraniferous and have moderate indoor radon associated with them. In particular, the Rincon shale may be the source of indoor radon levels greater than 4 pCi/L occurring in 75 percent of the homes in Santa Barbara County (Carlisle and Azzouz, 1991).

The Great Valley is made up of alluvium and colluvium derived from both the Coastal Ranges and the Sierra Nevada. Its radon potential is moderate overall but is controlled locally by source rock and permeability.

Regions 23, 24, 25: Columbia Plateau, Puget Lowland, Cascade Mountains, Northern Coastal Ranges, Klamath Mountains, and Willamette Valley

A comprehensive radon potential assessment of the Pacific Northwest has been done by Duval and others (1989b). The Columbia Plateau (Region 23) is underlain principally by Miocene basaltic and andesitic volcanic rocks, and tuffaceous sedimentary rocks and tuff. The soils formed from these rocks are low in uranium concentration, and indoor radon is generally low, giving the region an overall low radon potential. An extensive veneer of Pleistocene glaciofluvial outwash, eolian, and lacustrine deposits in the northern part of the Columbia Plateau (Region 24) contains locally highly permeable soils and relatively high soil uranium levels and has moderate radon potential. The subprovinces of the Blue Mountains and Joseph Upland in the central Columbia Plateau also include significant outcrop areas of Jurassic and Triassic sedimentary and volcanic rocks, weakly metamorphosed in many areas, and younger intrusive rocks that have a low to moderate radon potential.

The Puget Lowland in the northern part of Region 23 is underlain almost entirely by glacial deposits and Holocene alluvium. Most of the glacial and alluvial material of the Puget Lowland is derived from the Cascades to the east and the mountains of the Olympic peninsula to the west. The Puget Lowland overall has low radon potential because of high soil moisture and low uranium content of soils. Most townships from Tacoma northward have average indoor radon levels less than 1 pCi/L.

The Cascade Mountains (Region 23) extend from southwestern Oregon to northwestern California and can be divided into two geologic terranes: a northerly terrane composed principally of Mesozoic metamorphic rocks intruded by Mesozoic and Tertiary granitic rocks, and a southerly terrane composed of Tertiary and Holocene volcanic rocks that form locally thick volcanic ash deposits east of the Cascade Mountains. Overall, the sparsely populated Cascade Mountain Province has low radon potential because of the low uranium and high moisture contents of the soils.

The Coastal Range Province (Region 23) extends from the Olympic Peninsula of Washington south to the coastal parts of the Klamath Mountains in southwestern Oregon. In Washington, they are underlain principally by Cretaceous and Tertiary continental and marine sedimentary rocks and pre-Miocene volcanic rocks. In Oregon, the northern part of the Coastal Ranges are underlain principally by marine sedimentary rocks and mafic volcanic rocks of Tertiary age. The southern part of the Coast Range is underlain by Tertiary estuarine and marine sedimentary rocks, much of them feldspathic and micaceous. The Klamath Mountains are dominated by Triassic to Jurassic metamorphic, volcanic, and sedimentary rocks, with some Cretaceous intrusive rocks. These metamorphic and volcanic rocks are largely of mafic composition. Large masses of ultramafic rocks occur throughout the Klamath area. The radon potential of the Coastal Range Province is low overall. Most of the area has high rainfall and, as a consequence, high soil moisture. Uranium in the soils is typically low. Highly permeable, excessively well-drained soils may cause locally elevated indoor radon levels.

River alluvium and river terraces underlie most of the Willamette River Valley (Region 25); however, many of the hills that rise above the plains are underlain by Tertiary basalts and marine sediments. The Willamette River Valley has moderate radon potential overall. Much of the area has somewhat elevated uranium present in soils and many areas have excessively drained soils and soils with high emanating power. Many townships in the valley have indoor radon averages between 2 and 4 pCi/L.

Region 26: Hawaii

The volcanic island chain of Hawaii consists of Late Pliocene, Pleistocene, and Recent volcanic rock, predominantly basaltic lavas, ashes and tuffs, with minor carbonate and clastic marine sediments, alluvium, colluvium, dune sands, and mudflow deposits. Although some soils have soil-gas radon concentrations greater than 1,000 pCi/L (G. M. Reimer, written communication, 1991), suggesting that all structures within basements should be tested, the lifestyles of the inhabitants and typical local architecture contribute to the overall low radon potential of the state.

Regions 27, 28: Alaska

Alaska is divided from north to south into two main provinces: the Arctic Coastal Plain and the Northern Foothills comprise one province (Region 27), and the Arctic Mountains, the Central Province, and the Border Ranges comprise the other (Region 28). The Arctic coastal plain province (North Slope) consists primarily of Quaternary sedimentary rocks, mostly alluvium, glacial debris, and eolian sand and silt. A belt of Tertiary sedimentary rocks along the eastern third of the area separates the coastal plains from the foothills to the south. The Foothills province is largely composed of marine and nonmarine Cretaceous sandstone and shale, much of which is folded into westerly trending anticlines and synclines. This area has low radon potential.

The Arctic Mountains province is composed largely of faulted upper Precambrian and Paleozoic marine sedimentary rocks. The Central Province consists mostly of Precambrian and Paleozoic metamorphic rocks, Precambrian through Cretaceous mostly marine sedimentary rocks, Mesozoic intrusive and volcanic rocks, Tertiary and Quaternary mafic volcanic rocks, flat-lying Tertiary basin-fill (nonmarine clastic rocks), and Quaternary surficial deposits. The central province has several areas of uraniferous granites together with felsic intrusive and volcanic rocks. The schist that produces high indoor radon near Fairbanks is in this area.

The Border Ranges area includes the Alaska-Aleutian subprovince, the Coastal Trough Province, and the Pacific Border Ranges Province. The area is composed of several mountain belts separated by a series of depositional Cenozoic basins in a manner somewhat similar to that of the Basin and Range Province of the southwestern United States. Rocks exposed in the area include Paleozoic mafic volcanic rocks; Mesozoic mafic volcanic flows and tuffs, together with various units of shale, conglomerate, graywacke, and slate; and Tertiary and Quaternary intermediate volcanic rocks, Tertiary felsic intrusives, and Quaternary glacial deposits, eolian sand, and silt. The Coastal Trough Province contains thick sequences of Tertiary continental clastic and volcanic rocks penetrated by Tertiary intrusive rocks. Mesozoic sedimentary rocks and Pleistocene glacial deposits are abundant in some areas. Cretaceous and Jurassic sedimentary and metamorphic rocks, interbedded with mafic volcanic and intrusive rocks, comprise most of the Border Ranges. A fairly large area of lower Tertiary sedimentary and volcanic rocks is found in the Prince William Sound area. In much of this part of Alaska, annual rainfall is high (as much as 170 in), and water saturation likely retards gas flow in soils on all but the steepest of slopes. The Arctic Mountains, central Alaska, and Border Ranges area have an overall moderate radon potential.

ACKNOWLEDGMENTS

We thank M. Reimer, P. Nyberg, and R. Wanty for their review of the manuscript.

REFERENCES CITED

Archer, V. E., Gillam, J. D., and Wagoner, J. K., 1976, Respiratory disease mortality among uranium miners: Proceedings of the New York Academy of Science, v. 271, p. 280–293.

Carlisle, D., and Azzouz, H., 1991, Geological parameters in radon risk assessment—A case history of deliberate exploration, in Proceedings of the 1991 International Symposium on Radon and Radon Reduction Technology, Volume 5, Preprints: U.S. Environmental Protection Agency Report, Paper IX-6, 14 p.

Cohen, B. L., 1990, Surveys of radon levels in homes by University of Pittsburgh Radon Project, in Proceedings of the 1990 International Symposium on Radon and Radon Reduction Technology, Vol. 3, Preprints: U.S. Environmental Protection Agency Report EPA/600/9-90/005c, Paper IV-3, 17 p.

Cohen, B. L., and Gromicko, N., 1988, Variation of radon levels in U.S. homes with various factors: Journal of the Air Pollution Control Association, v. 38, p. 129–134.

Coveney, R. M., Jr., Hilpman, P. L., Allen, A. V., and Glascock, M. D., 1988, Radionuclides in Pennsylvanian black shales of the midwestern United States, in Marikos, M. A., and Hansman, R. H., eds., Geologic causes of natural radionuclide anomalies: Proceedings of the GEORAD Conference, Missouri Department of Natural Resources, Special Publication 4, p. 25–42.

Cross, F. T., 1987 effects, in Cothern, C. R., and Smith, J. E., Jr., eds., Environmental radon: New York, Plenum Press, p. 215–248.

De Voto, R. H., 1984, Uranium exploration, in DeVivo, B., Ippolito, F., Capaldi, G., and Simpson, P. R., eds., Uranium geochemistry, mineralogy, geology, exploration and resources: London, Institution of Mining and Metallurgy, p. 101–108.

——, 1988, Uranium geology and exploration: Golden, Colorado, Colorado School of Mines.

Durrance, E. M., 1986, Radioactivity in geology: Principles and applications: New York, Wiley & Sons, 441 p.

Duval, J. S., Cook, B. G., and Adams, J.A.S., 1971, Circle of investigation of an airborne gamma-ray spectrometer: Journal of Geophysical Research, v. 76, p. 8466–8470.

Duval, J. S., Jones, W. J., Riggle, F. R., and Pitkin, J. A., 1989a, Equivalent uranium map of the conterminous United States: U.S. Geological Survey Open-File Report 89-478, 10 p.

Duval, J. S., Otton, J. K., and Jones, W. J., 1989b, Estimation of radon potential in the Pacific Northwest using geological data: U.S. Department of Energy Bonneville Power Administration Report DOE/BP-1234, 146 p.

Dziuban, J. A., Clifford, M. A., White, S. B., Bergstein, J. W., and Alexander, B. V., 1990, Residential radon survey of twenty-three states, in Proceedings of the 1990 International Symposium on Radon and Radon Reduction Technology, Vol. III: Preprints. U.S. Environmental Protection Agency Report EPA/600/9-90/005c, Paper IV-2, 17 p.

Flood, R. L., Thomas, T. B., Suneson, N. H., and Luza, K. V., 1990, Radon potential map of Oklahoma: Oklahoma Geological Survey Map GM-32, scale 1:750,000.

Froelich, A. J., and Robinson, G. R., Jr., eds., 1988, Studies of the Early Mesozoic basins of the eastern United States: U.S. Geological Survey Bulletin 1776, 423 p.

Gates, A. E., and Gundersen, L.C.S., 1989, The role of ductile shearing in the concentration of radon in the Brookneal mylonite zone, Virginia: Geology, v. 17, p. 391–394.

Grasty, R. L., 1989, The relationship of geology and gamma-ray spectrometry to radon in homes: EOS Transactions of the American Geophysical Union, v. 70, p. 496.

Gundersen, L.C.S., 1989a, Predicting the occurrence of indoor radon: A geologic approach to a national problem: EOS Transactions of the American Geophysical Union, v. 70, p. 280.

——, 1989b, Anomalously high radon in shear zones, in Osborne, M., and Harrison, J., symposium cochairmen, Proceedings of the 1988 Symposium on Radon and Radon Reduction Technology, Volume 1, Oral presentations: U.S. Environmental Protection Agency Publication EPA/600/9-89/006A, p. 5-27 to 5-44.

——, 1989c, Geologic controls on radon: Geological Society of America Abstracts with Programs, v. 21, no. 2, p. 19–20.

——, 1991, Radon in sheared metamorphic and igneous rocks, in Gundersen, L.C.S., and Wanty, R. B., eds., Field studies of radon in rocks, soils, and water: U.S. Geological Survey Bulletin 1971, p. 39–50.

Gundersen, L.C.S., Reimer, G. M., and Agard, S. S., 1988a, Correlation between geology, radon in soil gas, and indoor radon in the Reading Prong, in Marikos, M. A., and Hansman, R. H., eds., Geologic causes of natural radionuclide anomalies: Missouri Department of Natural Resources Special Publication 4, p. 91–102.

Gundersen, L.C.S., Reimer, G. M., Wiggs, C. R., and Rice, C. A., 1988b, Map showing radon potential of rocks and soils in Montgomery County, Maryland: U.S. Geological Survey Miscellaneous Field Studies Map MF-2043, scale 1:62,500.

Hall, F. R., Boudette, E. L., and Olszewski, W. J., Jr., 1987, Geologic controls and radon occurrence in New England, in Graves, B., ed., Radon in ground water: Chelsea, Michigan, Lewis Publishers, p. 15–30.

Henry, M. E., Kaeding, M., and Montverde, D., 1989, Radon in soil gas and gamma ray activity measurements at Mulligan's Quarry, Clinton, New Jersey: Geological Society of America Abstracts with Programs, v. 21, p. 22.

Hopke, P. K., 1987, Radon and its decay products: Occurrence, properties, and health effects: American Chemical Society Symposium Series 331, 609 p.

Jenny, H., 1935, The clay content of the soil as related to climatic factors, particularly temperature: Soil Science, v. 40, p. 111–128.

Kisvarsanyi, E. B., 1988, Radioactive HHP (high heat production) granites in the Precambrian terrane of southeastern Missouri, in Marikos, M. A., and Hansman, R. H., eds., Geologic causes of natural radionuclide anomalies: Proceedings of the GEORAD Conference, Missouri Department of Natural Resources Special Publication 4, p. 5–15.

Klusman, R. W., and Jaacks, J. A., 1987, Environmental influences upon mercury, radon, and helium concentrations in soil gases at a site near Denver, Colorado: Journal of Geochemical Exploration, v. 27, p. 259–280.

Kovach, E. M., 1945, Meteorological influences upon the radon content of soil

gas: EOS Transactions of the American Geophysical Union, v. 26, p. 241–248.
Kunz, C., Laymon, C. A., and Parker, C., 1989, Gravelly soils and indoor radon, *in* Osborne, M., and Harrison, J., symposium cochairmen, Proceedings of the 1988 Symposium on Radon and Radon Reduction Technology, Volume 1, Oral presentations: U.S. Environmental Protection Agency Publication EPA/600/9-89/006A, p. 5-75–5-86.
Lawrence, E. P., Wanty, R. B., and Briggs, P. H., 1989, Hydrologic and geochemical processes governing distribution of U-238 series radionuclides in ground water near Conifer, CO: Geological Society of America Abstracts with Programs, v. 21, p. A144.
McLemore, V. T., and Hawley, J. W., 1988, Preliminary geologic evaluation of radon availability in New Mexico: New Mexico Bureau of Mines and Mineral Resources Open-File Report 345, 31 p.
Muessig, K., and Bell, C., 1988, Use of airborne radiometric data to direct testing for elevated indoor radon: Northeastern Environmental Science, v. 7, p. 45–51.
Nash, J. T., Granger, H. C., and Adams, S. S., 1981, Geology and concepts of genesis of important types of uranium deposits: Economic Geology, 75th Anniversary Volume, p. 63–116.
National Research Council, 1988, Health risks of radon and other internally deposited alpha emitters: Report of the Committee on the Biological Effects of Ionizing Radiation (BEIR IV), National Research Council: Washington, D.C., National Academy Press, 624 p.
——, 1990, Health effects of exposure to low levels of ionizing radiation: Report of the Committee on the Biological Effects of Ionizing Radiation (BEIR V), National Research Council: Washington, D.C., National Academy Press, 436 p.
——, 1991, Comparative dosimetry of radon in mines and homes: Washington, D.C., National Academy Press, 256 p.
Nichols, C. E., 1984, Uranium exploration techniques, *in* DeVivo, B., Ippolito, F., Capaldi, G., and Simpson, P. R., eds., Uranium geochemistry, mineralogy, geology, exploration and resources: London, Institution of Mining and Metallurgy, p. 23–42.
Ogden, A. E., Welling, W. B., Funderburg, D., and Boschult, L. C., 1987, A preliminary assessment of factors affecting radon levels in Idaho, *in* Graves, B., ed., Radon in ground water: Chelsea, Michigan, Lewis Publishers, p. 83–96.
Otton, J. K., 1992, Radon in soil gas and soil radioactivity in Prince Georges County, Maryland: U.S. Geological Survey Open File Report no. 92-11, 18 p.
Otton, J. K., Schumann, R. R., Owen, D. E., and Chleborad, A. F., 1988, Geologic assessments of radon hazards: A Colorado case history, *in* Marikos, M. A., and Hansman, R. H., eds., Geologic causes of natural radionuclide anomalies: Proceedings of the GEORAD Conference, Missouri Department of Natural Resources Special Publication 4, p. 167.
Peake, R. T., and Gundersen, L.C.S., 1989, An assessment of radon potential of the U.S. Coastal Plain: Geological Society of America Abstracts with Programs, v. 21, p. 58.
Peake, R. T., Gundersen, L.C.S., and Wiggs, C. R., 1988, The Coastal Plain of the eastern and southern United States—An area of low radon potential: Geological Society of America Abstracts with Programs, v. 20, p. A337.
Reesman, A. L., 1988, Geomorphic and geochemical enhancement of radon emission in middle Tennessee, *in* Marikos, M. A., and Hansman, R. H., eds., Geologic causes of natural radionuclide anomalies: Proceedings of the GEORAD Conference, Missouri Department of Natural Resources Special Publication 4, p. 119–130.
Ronca-Battista, M., Moon, M., Bergsten, J., White, S. B., Holt, N., and Alexander, B., 1988, Radon-222 concentrations in the United States—Results of sample surveys in five states: Radiation Protection Dosimetry, v. 24, p. 307–312.
Rose, A. W., Washington, J. W., and Greeman, D. J., 1988, Variability of radon with depth and season in a central Pennsylvania soil developed on limestone: Northeastern Environmental Science, v. 7, p. 35–39.
Schery, S. D., Gaeddert, D. H., and Wilkening, M. H., 1984, Factors affecting exhalation of radon from a gravelly sandy loam: Journal of Geophysical Research, v. 89, p. 7299–7309.
Schultz, A., and Brower, S. D., 1991, Geologic and environmental implications of high radon soil-gas concentrations in carbonate rocks of the Appalachian Great Valley, West Virginia: Geological Society of America Abstracts with Programs, v. 23, p. 125.
Schultz, A. P., and Wiggs, C., 1989, Preliminary results of a radon study across the Great Valley of West Virginia: Geological Society of America Abstracts with Programs, v. 21, p. 65.
Schumann, R. R., and Owen, D. E., 1988, Relationships between geology, equivalent uranium concentration, and radon in soil gas, Fairfax County, Virginia: U.S. Geological Survey Open-File Report 88-18, 28 p.
Schumann, R. R., Owen, D. E., and Asher-Bolinder, S., 1989a, Weather factors affecting soil-gas radon concentrations at a single site in the semiarid western U.S., *in* Osborne, M. C., and Harrison, J., eds., Proceedings of the 1988 EPA Symposium on Radon and Radon Reduction Technology, Volume 2, Poster presentations: U.S. Environmental Protection Agency Report EPA/600/9-89/006B, p. 3-1–3-13.
Schumann, R. R., Gundersen, L.C.S., Asher-Bolinder, S., and Owen, D. E., 1989b, Anomalous radon levels in crystalline rocks near Conifer, Colorado: Geological Society of America Abstracts with Programs, v. 21, p. A144–A145.
Schumann, R. R., Peake, R. T., Schmidt, K. M., and Owen, D. E., 1990, Correlations of soil-gas and indoor radon with geology in glacially derived soils of the northern Great Plains, *in* Proceedings of the 1990 EPA International Symposium on Radon and Radon Reduction Technology, Volume III: Preprints: U.S. Environmental Protection Agency Report EPA/600/9-90/005c, Paper VI-3, 14 p.
Sextro, R. G., Moed, B. A., Nazaroff, W. W., Revzan, K. L., and Nero, A. V., 1987, Investigations of soil as a source of indoor radon, *in* Hopke, P. K., ed., Radon and its decay products: American Chemical Society Symposium Series 331, p. 10–29.
Smith, D. K., 1984, Uranium mineralogy, *in* DeVivo, B., Ippolito, F., Capaldi, G., and Simpson, P. R., eds., Uranium geochemistry, mineralogy, geology, exploration and resources: London, Institution of Mining and Metallurgy, p. 43–88.
Smith, D. L., and Hansen, J. K., 1989, Distribution of potentially elevated radon levels in Florida based on surficial geology: Southeastern Geology, v. 30, p. 49–58.
Spitz, H. B., Wrenn, M. E., and Cohen, N., 1980, Diurnal variation of radon measured indoors and outdoors in Grand Junction, Colorado, and Teaneck, New Jersey, and the influence that ventilation has on the buildup of radon indoors, *in* Gesell, T. F., and Lowder, W. M., eds., The natural radiation environment, III: Springfield, Virginia, U.S. Department of Energy Report CONF-780422, v. 2, p. 1308–1329.
Sprinkel, D. A., 1988, Assessing the radon hazard in Utah: Utah Geological and Mineral Survey, Survey Notes, v. 22, p. 3–13.
Steinhäusler, F., 1988, Epidemiological evidence of radon-induced health risks, *in* Nazaroff, W. W., and Nero, A. V., eds., Radon and its decay products in indoor air: New York, John Wiley & Sons, p. 311–371.
Szabo, Z., and Zapecza, O. S., 1991, Geologic and geochemical factors controlling uranium, radium-226, and radon-222 in ground water, Newark Basin, New Jersey, *in* Gundersen, L.C.S., and Wanty, R. B., eds., Field studies of radon in rocks, soils, and water: U.S. Geological Survey Bulletin 1971, p. 243–265.
Tanner, A. B., 1964, Radon migration in the ground: A review, *in* Adams, J.A.S., and Lowder, W. M., eds., The natural radiation environment: Chicago, University of Chicago Press, p. 161–190.
——, 1980, Radon migration in the ground: A supplementary review, *in* Gesell, T. F., and Lowder, W. M., eds., Natural radiation environment III: Symposium Proceedings, v. 1, p. 5–56.
Turner-Peterson, C. E., 1988, A comparison of uranium-bearing sequences in the Newark Basin, Pennsylvania and New Jersey, and the San Juan Basin, New Mexico, *in* Froelich, A. J., and Robinson, G. R., Jr., eds., Studies of the

Early Mesozoic basins of the eastern United States: U.S. Geological Survey Bulletin 1776, p. 350–356.

U.S. Department of Energy, 1976, National Uranium Resource Evaluation preliminary report, Prepared by the U.S. Energy Research and Development Administration, Grand Junction, Colorado, GJO-11(76).

U.S. Geological Survey, 1992, Geologic Radon Potential of EPA Regions 1-10: U.S. Geological Survey Open-File Reports (in press).

U.S. Soil Conservation Service, 1987: U.S. Geological Survey National Atlas Sheet 38077-BE-NA-07M-00, scale 1:7,500,000.

Young, R. G., 1984, Uranium deposits of the world, excluding Europe, *in* De-Vivo, B., Ippolito, F., Capaldi, G., and Simpson, P. R., eds., Uranium geochemistry, mineralogy, geology, exploration and resources: London, Institution of Mining and Metallurgy, p. 117–139.

Wanty, R. B., and Schoen, R., 1991, A review of the chemical processes affecting the mobility of radionuclides in natural waters, with applications, *in* Gundersen, L.C.S., and Wanty, R. B., eds., Field studies of radon in rocks, soils, and water: U.S. Geological Survey Bulletin 1971, p. 183–194.

White, S. B., Bergsten, J. W., Alexander, B. V., and Ronca-Battista, M., 1989, Multi-state surveys of indoor Rn-222: Health Physics, v. 57, p. 891–896.

MANUSCRIPT ACCEPTED BY THE SOCIETY APRIL 6, 1992

Sensitivity of soil radon to geology and the distribution of radon and uranium in the Hylas zone area, Virginia

Alexander E. Gates
Department of Geology, Rutgers University, Newark, New Jersey 07102
Linda C. S. Gundersen
U.S. Geological Survey, Box 25046, Federal Center, MS-939, Denver, Colorado 80225

ABSTRACT

Shear zones and bedrock units as thin as 3 to 4 m in width can be detected using radon and equivalent uranium (eU) concentrations in soils overlying the Hylas zone area, Virginia. Soil radon concentrations in the area range from 144 to 12,081 pCi/L, with distinct contrasts in concentrations among many of the rock types. Equivalent uranium concentrations range from 0.4 to 9.8 ppm; they also reflect bedrock geology but less distinctly than soil radon, which is subject to surficial mechanical and chemical processes. The formation of saprolite and subsequent soil processes produce variability in uranium concentrations and consequently contacts and other fine geologic details can be obscured near the surface. Soil radon analyses at a depth of 75 cm, and using small (20 ml) samples, largely circumvent these problems and can be used as an effective geologic mapping tool in covered areas, provided that (1) there is a uranium concentration contrast in the bedrock, and that (2) the soil is derived directly from bedrock. The scale of variability of radon in soil gas indicates that radon potential maps should be constructed on the basis of geology rather than political divisions.

INTRODUCTION

Surveys of the distribution of radon (Rn-222) concentrations in soil gas have been a standard practice in uranium exploration since the late 1950s (Peacock and Williamson, 1962). Because radon is the daughter product of uranium, relatively high concentrations of radon indicate the presence of uranium ore bodies. Many uranium geologists consider radon surveying the most reliable of the geochemical exploration techniques and credit radon for discoveries of ore bodies at depths as much as 200 m (Gingrich and Fisher, 1976; Smith and others, 1976). Correlation of radon with small-scale features is less certain. Severne (1978) found correlations between radon and uranium concentrations at the scale of 1 to 2 m whereas Czarnecki and and Pacer (1983) found that correlations were on the scale of 100 or 200 m.

Environmental radon appears also to be largely controlled by local geology. Small but distinct geologic features appear to be responsible for many anomalous indoor radon occurrences in the eastern United States (Gundersen and others, this volume). Features such as shear zones (Gundersen and others, 1988; Gates and Gundersen, 1989; Gates and others, 1990), pegmatites (Grauch and Zarinski, 1976), locally deep soils developed on karst features (Schultz and Wiggs, this volume), certain stratigraphic contacts (Gundersen and others, 1988; Hand and Banikowski, 1988), and localized layers such as glauconitic sediments (Gundersen and Peake, this volume)—although areally limited—are often enriched in uranium and are more likely to produce anomalous radon than are large-scale features.

In spite of the results of uranium exploration and environmental radon research, ongoing government assessments of radon potential are based on inexact correlations of indoor radon concentrations with the geology of an area using zip code or county boundaries. These assessments are being used to legislate building and home testing procedures. Because these zip code or county areas may contain several geologic features of varying radon

Gates, A. E., and Gundersen, L.C.S., 1992, Sensitivity of soil radon to geology and the distribution of radon and uranium in the Hylas zone area, Virginia, *in* Gates, A. E., and Gundersen, L.C.S., eds., Geologic Controls on Radon: Boulder, Colorado, Geological Society of America Special Paper 271.

potential, such correlations may lead to erroneous oversimplifications. The resulting radon potential maps are almost invariably inapplicable to single houses or neighborhoods.

In this study, the limits of resolution for the correlation of soil radon concentrations with geology are qualitatively evaluated in a complex terrane. The Grenville Goochland terrane, Virginia (Fig. 1), has a high radon potential based on preliminary geologic and geochemical studies (Baillieul and Dexter, 1982; Gates and others, 1990). The present study demonstrates that most of the study area has a low radon potential, but that specific structures and areally limited units have the potential to produce anomalous radon concentrations. Because the rock units of the Goochland terrane have strongly contrasting uranium and radon concentrations, soil radon signatures can be identified for each unit. Using these signatures, rock units are shown to be traceable through coeval areas based on radon surveys. We therefore propose that soil radon surveying can be a powerful geologic mapping tool, and we offer a new, finer scale for radon potential evaluation based on the size of the radon-producing feature.

STRATIGRAPHY

The general geology and structure of the Goochland terrane is summarized by Farrar (1984) and in unpublished masters' theses (Bobyarchick, 1976; Bourland, 1976; Poland, 1976; Reilly, 1980). In this chapter, the only rock units considered are those for which radon and uranium concentrations were analyzed. The stratigraphy for the metamorphic units is based on structural relations because no sedimentary facing data exists. The major rock units and their subdivisions are summarized in Table 1 and include the Grenville-age State Farm Gneiss, Sabot Amphibolite, and Maidens Gneiss, and several intrusive units of various ages.

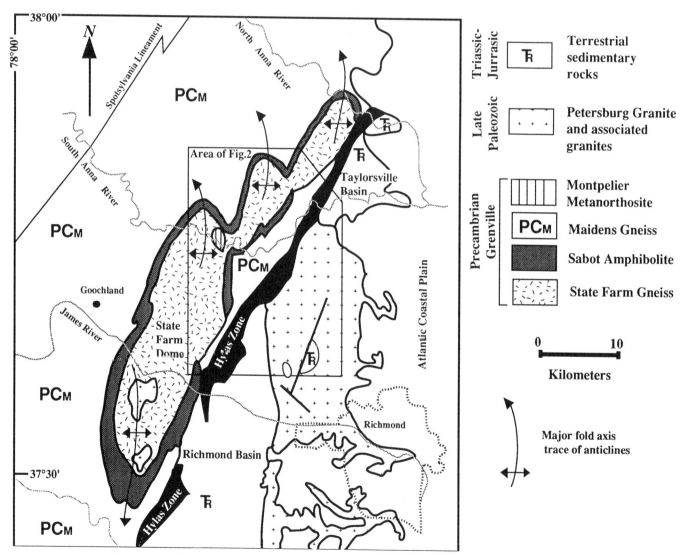

Figure 1. Geologic map of the Goochland terrane, eastern Piedmont Virginia (modified from Farrar, 1984), showing location of study area (Fig. 2).

TABLE 1. STRATIGRAPHY OF THE HYLAS
ZONE AREA, VIRGINIA

	Petersburg Granite	Coarse-grained granite, foliated to mylonitic with K-spar, quartz, plagioclase, biotite, muscovite, apatite, zircon, allanite, opaques
	Montpelier Anorthosite	Coarse-grained, antiperthitic plagioclase with quartz, apatite, rutile, ilmenite, sphene, diopside, and foliated, medium-grained anorthosite with hornblende, biotite, and muscovite
Maidens Gneiss	Granulite Gneiss (west)	Medium-grained, orthopyroxene, clinopyroxene, plagioclase, garnet with ilmenite, rutile, quartz, K-spar, hornblende, and biotite
Maidens Gneiss	Granitic Gneiss (east)	Foliated to mylonitic, medium-grained hornblende granitic gneiss with K-spar, plagioclase, quartz, hornblende, garnet, apatite, muscovite and opaques, and hornblende-biotite granitic gneiss with biotite and excluding garnet
Maidens Gneiss	Quartzite	95 to 98% quartz with biotite and opaques
Maidens Gneiss	Calc-silicate	Scapolite-diopside gneiss with plagioclase, sphene, epidote, tremolite, phlogopite, K-spar, clinozoisite, clacite, zircon, apatite, opaques
Maidens Gneiss	Granite pods	Foliated to mylonitic, medium-grained alkali-feldspar granite to monzogranite with K-spar, quartz, plagioclase, biotite, muscovite, garnet, chlorite, epidote, apatite, opaques
Maidens Gneiss	Amphibolite	Medium-grained hornblende, andesine, clinopyroxene, sphene, apatite, magnetite
Maidens Gneiss	Aluminous Gneiss	Biotite-garnet migmatitic gneiss with muscovite, quartz, plagioclase, kyanite, staurolite, K-spar, sillimanite, apatite, rutile, zircon, opaques, contains K-spar rich leucosome with quartz, plagioclase, biotite, kyanite, allanite, episote, and zircon
Sabot	Amphibolite	Hornblende-andesine gneiss with clinopyroxene cores in hornblende, accessory sphene, apatite, magnetite, ilmenite
State Farm Gneiss	Tonalitic Gneiss	Tonalitic to hornblende dioritic gneiss with plagioclase, quartz, biotite, K-spar, hornblende, clinopyroxene, sphene, apatite, rutile, zircon, opaques, locally gradational into Sabot amphibolite
State Farm Gneiss	Garnet-biotite Gneiss	Garnets (2 to 3 cm in diameter) in a matrix of quartz, plagioclase muscovite, K-spar and kyanite with apatite, zircon, and opaques
State Farm Gneiss	Granitic Gneiss	Granitic to granodioritic gneiss with K-spar, quartz, plagioclase, biotite, hornblende, apatite, zircon, and opaques

Mesozoic diabase dikes and terrestrial siliciclastics also occur in the area.

The State Farm Gneiss (Brown, 1937; Poland, 1976; Reilly, 1980) is structurally the lowest unit in the Goochland terrane and comprises the cores of the en echelon domes (Fig. 1). It is divided into several compositional subunits, including tonalite, garnet biotite gneiss, and granite gneiss. Samples of the granite gneiss have been dated at $1,031 \pm 94$ Ma, using whole-rock Rb/Sr methods (Glover and others, 1978, 1982). The overlying Sabot Amphibolite is in sharp contact with the State Farm Gneiss in most of the map area but appears gradational in the north. It is lithologically and stratigraphically similar to other amphibolites in the lower Maidens Gneiss.

Both the Sabot Amphibolite and the thick amphibolite in the Maidens Gneiss are overlain by 3- to 8-m-thick massive quartzite units. The quartzites grade upward into migmatitic garnet-biotite gneisses that are interpreted as metapelites. The metapelites are kyanite and staurolite bearing near their contact with the Hylas zone and contain granular sillimanite in all other areas. A calc-silicate gneiss also occurs discontinuously interlayered with the metapelite. Pods, lenses, and a single 10-m-thick layer of fine-grained, quartzofeldspathic gneiss with minor garnet and muscovite occur within the middle and upper portions of the Maidens Gneiss. An extensive quartzofeldspathic gneiss unit forms the majority of the upper Maidens Gneiss. The undifferentiated units to the northwest of the domes (Fig. 1), to which this unit apparently correlates, exhibit granulite facies assemblages (Farrar, 1984).

The Petersburg Granite is a large pluton that intruded the eastern Goochland terrane at 330 ± 8 Ma (Wright and others, 1975). The granite is dominantly undeformed but locally sheared, forming a fine-grained, well-foliated mylonitic gneiss. Proposed related plutons include the Fine Creek Mills granite (Poland, 1976; Reilly, 1980) and the Flat Rock granite (Reilly, 1980).

Granitic pegmatite dikes of unknown age intrude the Goochland terrane and are dominantly 0.5 to 2 m thick but may be as thick as 200 m (Bobyarchick, 1976). The highly radioactive pegmatites are easily located with a gamma scintillometer and contain abundant potassium feldspar with quartz, muscovite, biotite, apatite, and zircon. The pegmatite dikes are concentrated along the Hylas zone; because they have suffered pre-Alleghanian deformation, their presence may indicate an earlier period of movement along the zone. They may also be related to Petersburg plutonism.

Siliciclastic sedimentary rocks of Triassic-Jurassic age are deposited in the Richmond and Taylorsville rift basins. The Hylas zone was reactivated as the basin-bounding normal fault for both basins. The sedimentary rocks are primarily oxidized sandstones and conglomerates, but gray sandstones with coal seams occur locally within the basins. All basal rocks and rocks in the western portions of each basin are dominated by pebble, cobble, and locally boulder conglomerates. The clasts consist of all lithologies of the Goochland terrane but are commonly mylonite and Petersburg Granite (Weems, 1974).

STRUCTURE

An intense deformational event is associated with Grenville-age granulite facies metamorphism of the State Farm Gneiss, Maidens Gneiss, and Sabot Amphibolite. This event (Farrar, 1984) produced a penetrative foliation and recumbent isoclinal, intrafolial folds that are well displayed in the State Farm and Maidens Gneisses.

A second major deformational event associated with amphibolite facies metamorphism is attributed to the Late Paleozoic Alleghanian orogeny (Farrar, 1984; Gates and Glover, 1989). Structures formed during this deformation include the State Farm and related northeast-trending en echelon domes (Gates and Glover, 1989) (Fig. 1). The domes form a major northeast-trending antiform that is bounded to the east by the Hylas zone. Each dome has a north-northwest- to northeast-trending fold axis that curves into the Hylas zone. The domes are slightly asymmetric to the east. Many macroscopic and small megascopic asymmetric folds occur in the area. The asymmetry of all folds show a consistent clockwise vergence sense, and the folds appear intimately related to both major and minor shear zones.

Shear zones

The most prominent structural feature in the study area is the Hylas shear zone. The Hylas zone is 0.5 to 2.4 km wide, northeast-striking, and displays several periods of movement (Bobyarchick and Glover, 1979). It contains mylonitic and cataclastic Petersburg Granite and units of the middle and upper Maidens Gneiss. Several northeast-striking splays of the Hylas zone were identified, as were minor northwest-trending shear zones with prominent sinistral transcurrent drag (Fig. 2).

Deformation in the Hylas zone and smaller shear zones are divided into three groups: ductile, brittle-ductile transition, and brittle. Commonly, a single zone will record all three behaviors. Minerals in the rocks that record deformation in the ductile regime exhibit synkinematic recrystallization and flow. Ductile deformation in the Hylas zone consistently indicate dextral strike-slip (Gates and Glover, 1989) and produce C and S bands (Berthe and others, 1979). The S bands define the long axis of the finite strain ellipsoid; the C bands define the shear plane (Simpson and Schmid, 1983). The initial angle between the two is 45° but decreases as S is rotated into C. Assuming simple shear, the angle between the two can be used to calculate shear strain; the smaller the angle, the greater the shear. Ductile deformation also produced rotated mica and feldspar porphyroclasts and quartz ribbons in the Petersburg Granite, and shear bands (Platt and Vissers, 1980), rotated porphyroclasts, and quartz ribbons in the Maidens pelitic gneiss.

At the brittle-ductile transition, some minerals exhibit fracturing and cataclasis. Zones that exhibit semi-brittle deformation are also dextral, occur as thin bands or localized areas within the shear zones and postdate the ductile deformation. Deformation of the Petersburg Granite at the brittle-ductile transition is characterized by brittle feldspar and accessory minerals except plastic quartz and micas (Gates and Glover, 1989).

Brittle deformation clearly postdates all folding and brittle-ductile faulting. Brittle faulting is primarily normal but strike-slip displacement is also common (Bobyarchick, 1976). The brittle faults are single slickensided planes or zones of cataclasite or gouge. Cataclasite and gouge are commonly mineralized with chlorite, epidote, calcite, and zeolites (Bobyarchick and Glover, 1979).

URANIUM AND RADON

Methodology

Soil radon and equivalent uranium (eU) concentrations in soil were measured in the Goochland terrane in the vicinity of the Hylas zone (Fig. 2). The area covered by the survey is approximately 200 km^2. A standard grid pattern of sampling was employed for part of the eU survey. Considering, however, that some of the geologic features are tens of meters wide, most eU and radon sample locations were chosen to best cover the maximum number of units and structural features. Short traverses of closely spaced samples were made perpendicular to strike in geologically critical areas. The length of the traverses varied from 0.75 to 10 km, according to the size of the feature. Spacing of samples was also chosen to best cover the feature and was usually >150 m. Over thin units and mylonites, the spacing was usually <100 m. Fifty-seven samples were taken adjacent to rock exposures to ensure proper correlation of radon and eU concentrations with rock types. The other samples were taken in covered areas and were correlated to rock types based on structural extrapolation from outcrops.

Equivalent uranium concentration (eU) of soils was measured at 57 radon sample locations, as well as at spaced intervals throughout the study area for a total of 173 measurements. Equivalent uranium was measured using a multi-channel calibrated gamma scintillometer and 5-min counting intervals at each location. Counts were converted to eU in parts per million using the machine calibration.

Radon in soil gas was determined using the methods described in Gates and Gundersen (1989) and Reimer (1992). Soil gas samples of 20 cm^3 were extracted from soil at a depth of 75 cm, which corresponds to the B soil horizon in the study area. Because samples are small and extracted at low vacuum, they reflect only the composition of the soil gas at the sampling point. Because they are grab samples, the radon concentrations are applicable to soil conditions at the time of sampling. Radon concentration, expressed in picocuries per liter, was determined using an alpha scintillometer several hours after sampling to eliminate Rn-220 and for each sample to equilibrate. In this area, the soils at the sampling depth are locally derived, relatively clay-rich, and of low to moderate permeability. Most soils have a well-developed, clay-rich B layer and are therefore not subject to appreciable atmospheric dilution. Duplicate samples were taken and every tenth sample analyzed for quality assurance. Most sampling was done in June 1988 (115 samples including duplicates); we also sampled in May 1990 (35 samples including duplicates). Both time periods exhibited hot, dry, stable weather and the measurements are considered comparable.

Uranium distribution

The eU distribution shows high contrast for specific units in the Hylas zone area, but each unit shows great variability in concentrations and nearly all overlap (Table 2; Figs. 2a, 3a). The pegmatites, Petersburg Granite, mylonite, Triassic conglomerate, and Triassic sandstones show relatively high average eU (3.9 to 6 ppm) but with variations of ±1 to ±3.5 ppm. The pegmatites exhibit higher eU contents over thicker dikes. The Triassic sandstones contain generally lower eU concentrations than the conglomerates. The Maidens Gneiss metapelite unit has a moderate average eU concentration but shows a similar range of eU concentrations as the Petersburg Granite. The State Farm granitic, tonalitic, and garnet-biotite gneisses, the Maidens granitic gneiss, the Sabot Amphibolite, and Maidens Gneiss amphibolites exhibit the lowest eU concentrations in the area. All eU concentrations are <4 ppm, and all averages are <3 ppm with variability of ±0.8 to ±1.5 ppm.

The eU concentration of soil over mylonite depends largely on the protolith of the mylonite and therefore although elevated, the eU is highly variable. The eU concentration of the Hylas zone is clearly elevated in comparison to the other Goochland terrane gneisses (Fig. 3a).

Radon in soil gas distribution

Radon concentrations in soil exhibit a strong dependence on geology (Table 2; Figs. 2b, 3b). Signatures of soil radon for each unit are based on two parameters: the average concentration and the range of concentrations within the unit. The units generally fall into one of two groups. Soils formed from the State Farm Gneiss, Maidens Gneiss, and Sabot Amphibolite generally contained <1,800 pCi/L and soils from the Petersburg Granite, pegmatites, mylonites, and Triassic units generally contained >1,800 pCi/L.

There is considerable overlap between the units within each group but some generalities can be observed. In the <1,800 pCi/L group, the State Farm garnet-biotite gneiss and the Maidens hornblende-biotite gneiss had higher radon contents (>~900 pCi/L) than the Sabot and Maidens Gneiss amphibolites (<~900 pCi/L). In the >1,800 pCi/L group, the pegmatites and Triassic conglomerate contained >4,200 pCi/L, whereas the Petersburg Granite contained largely <3,100 pCi/L.

RADON AS A MAPPING TOOL

There is a significant overlap of soil radon and especially eU values among the geologic units in the Hylas zone area. Radon concentrations, however, can be used to discern certain units in covered areas depending upon the adjacent unit. A unit in the >1,800 pCi/L group is easily distinguished from a unit in the <1,800 pCi/L group. To discriminate units with overlapping radon concentrations, both radon variability and geologic information about the specific unit must be considered.

The unit with one of the largest variability of radon concentrations is mylonite. The radon concentrations of mylonite overlap all units in the >1,800 pCi/L group and even a few data in the 1,800 pCi/L group. If considered in terms of protolith and amount of strain, however, the results are much clearer. Radon in soil above mylonites was consistently higher than above undeformed counterparts by as much as 11-fold (Fig. 4a). Radon concentrations above the Hylas zone were consistently the highest in the study area (as much as 12,051 pCi/L). As with the Brookneal zone (Gates and Gundersen, 1989), radon over mylonitic Petersburg Granite varies directly with the angle between C and S bands (Fig. 4b). These angles directly reflect shear strain in mylonites (see Ramsay and Graham, 1970). Samples of soil gas were taken adjacent to each mylonite outcrop where C/S angles were measured for direct comparison.

The graph of radon vs. C/S angle for the Hylas zone has a curve remarkably similar to that for the Brookneal mylonite zone, Virginia (see Fig. 5 in Gates and Gundersen, 1989). The Brookneal zone is 6 km wide and strain varies gradually across it. The Hylas zone, on the other hand, is only 0.7 km wide where the radon and shear strain were investigated. Both radon and shear strain varied markedly across this narrow traverse and yet correlate as well as the Brookneal zone. This sensitivity indicates that, despite the large radon concentration gradients in the soil, radon diffusion is apparently slow or local such that equilibrium is maintained on a small scale.

Soils above mylonitic Maidens Gneiss also showed elevated radon concentrations relative to undeformed counterparts (Fig. 4a). Therefore, unless the mylonite zone is completely contained within a single unit, it should be identifiable on map pattern as a band of greatly elevated radon that persists across an area regardless of the host unit.

Using conventional geologic mapping, there is a strong potential to misidentify units in covered areas. By considering the petrology and chemical composition of a rock in conjunction with measured soil radon of adjacent or related units, better discernment of units is possible. Using these results, map patterns can be reinterpreted in covered areas. The State Farm granite and tonalite gneisses show bimodal distributions of radon, a group with <1,000 pCi/L and a group with >1,000 pCi/L. The State Farm garnet-biotite gneiss contains a single group that is very similar to the >1,000 pCi/L group of the granite gneiss. The >1,000 pCi/L group may reflect the occurrence of mylonite in either group or a compositional change within the rock unit. The map pattern of the garnet-biotite gneiss may be more complex than could be resolved using standard geologic mapping. Systematic radon sampling could resolve details in the State Farm Gneiss stratigraphy.

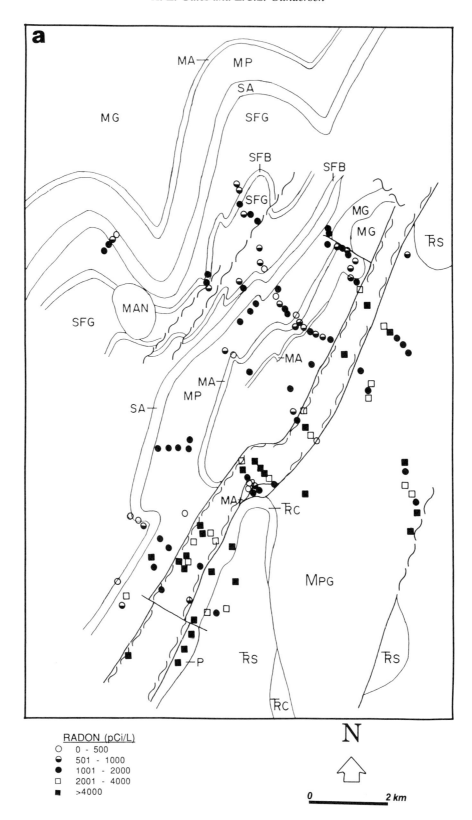

Figure 2. Geologic map of the Hylas zone area (after Gates and Glover, 1989). a, Soil radon; b, equivalent uranium. State Farm Gneiss includes lower granitic gneiss that is gradational into upper tonalitic gneiss (SFG); both contain metapelite (SFB). Montpelier anorthosite (MAN) intrudes State

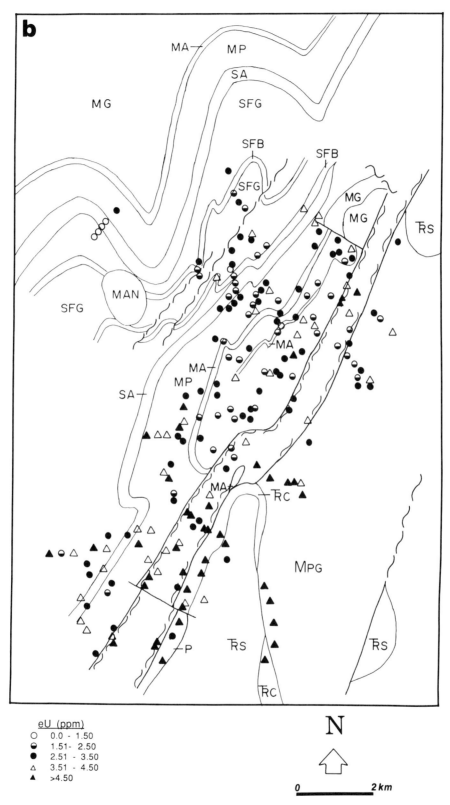

Farm Gneiss. Maidens Gneiss is undifferentiated except for major, continuous amphibolite layers (MA), and the Sabot Amphibolite (SA). All metapelites (MP) between the amphibolites are succeeded by granitic and granulite gneiss (MG). Pegmatites = P; Petersburg Granite = MPG; Triassic conglomerate = TRC; Triassic sandstone = TRS.

TABLE 2. RADON AND EQUIVALENT URANIUM DATA FOR THE HYLAS ZONE AREA, VIRGINIA

Unit	Radon (pCi/L)			Equivalent Uranium (ppm)		
	Number	Average	Range	Number	Average	Range
S.F. Granite Gneiss*	14	1,009	467 to 1,954	15	2.33	0.98 to 3.87
S.F. Garnet-Biotite Gneiss*	3	1.577	1,354 to 1,996	6	2.97	2.04 to 3.95
Amphibolites†	13	530	144 to 869	15	2.10	0.60 to 3.35
Maidens Metapelite	22	1,245	388 to 2,585	43	3.42	1.87 to 5.41
Maidens Granite Gneiss	14	1,170	801 to 1,545	28	2.89	1.86 to 4.77
Pegmatite	3	4,944	4,208 to 6,133	4	6.05	3.38 to 9.67
Petersburg Granite	9	2,633	1,344 to 6,223	14	4.38	2.31 to 8.81
Mylonite (all)	31	4,574	1,502 to 12,081	37	4.53	1.84 to 7.10
Triassic Conglomerate	3	6,147	4,504 to 9,398	8	5.94	4.29 to 7.99
Triassic Sandstone	4	3,565	1,497 to 6,611	3	3.90	3.16 to 4.87

*S.F. = State Farm.
†Amphibolites = includes both Sabot and Maidens Gneiss.

The Maidens Gneiss metapelite shows a surprising variability in radon concentrations. The metapelite is relatively thick and therefore not as prone to errors from mislocation of contacts in covered areas as are thinner units. The metapelite, however, is commonly migmatitic and includes small granite and pegmatite bodies. First melts are commonly enriched in incompatible elements, including uranium, relative to the parent rock. The pegmatites and leucosome of the migmatites may have been enriched in uranium and the melanosome depleted in uranium during melting. Radon in soils over the pegmatites in the metapelite was not discriminated and the radon contribution from the melanosome and leucosome could not be discriminated on this scale of investigation. Tightly spaced (on a scale of centimeters to meters) sampling might resolve the relative contributions. An alternate explanation to the radon variability in the Maidens Gneiss metapelite is unrecognized lithologies within the unit. Thin layers of amphibolite, quartzite, or calc-silicate might cause the same variability in soil radon. As in the State Farm Gneiss, systematic radon sampling in conjunction with detailed geochemical studies could resolve the details.

The Triassic sandstones and conglomerates are composed primarily of detrital quartz and feldspar. They are also rich in rock fragments that are dominated by Hylas zone mylonite and Petersburg Granite. The Triassic sediments are therefore enriched in uranium by virtue of the uranium content of their constituents. In addition, Baillieul and Dexter (1982) found that coal seams and some gray sandstones are enriched in uranium.

The variability of the eU within each unit is high, resulting in extensive overlap of concentrations among the units. In the 57 samples in which soil radon and eU were measured simultaneously, there is a diffuse positive correlation between soil radon and soil eU (Fig. 5a). However, two distinct populations can be observed; all eU values greater than 4 ppm have radon greater than 1,000 pCi/L. Uranium and radium concentration is strongly affected by vegetation, human activity, and surficial processes. Radium is known to concentrate in surface vegetation, and depending on the composition and pH of surface waters, both uranium and radium may be soluble in the soil. The more mature and deeper the soil, the more likely radionuclides have been redistributed within the soil column. Human activity such as additions of phosphate fertilizers, or radioactive waste, landscaping, or cut and fill operations can also radically change the eU content of soils.

Contacts between units are not sharply defined by eU concentrations in the study area. Even at contacts between units with marked soil radon concentration contrasts, eU shows a gradual transition between the two (Fig. 5b). Near-surface processes may have effectively mixed the soils across contacts, resulting in gradual changes in eU concentrations. The map distribution of eU does reflect the general geology; however, the unit by unit analysis, as presented here, shows a less distinct correlation.

DISCUSSION

Comparison of equivalent uranium in soil and soil radon concentrations with bedrock geology in the Hylas zone area documents the variability of radon and uranium on several scales. Lithologic and tectonic contacts are discernable by variations in soil radon and eU concentrations along the cross-strike traverses. Units as thin as tens of meters can be tentatively identified on the map scale using eU, but soil radon appears to be a much more effective indicator of lithologic differences. Thin mafic units were readily apparent because they are distinctly lower in uranium (low eU) and produce less radon than surrounding units. Other thin units such as the pegmatites and Triassic conglomerate are also easily discernible because they contain more uranium and produce more radon than the surrounding units.

Because soil radon concentrations reflect bedrock so well, they can be used as mapping tools to augment geologic data. As a result of the anomalously high soil radon levels discovered within the Petersburg Granite, a previously unrecognized zone of proto-

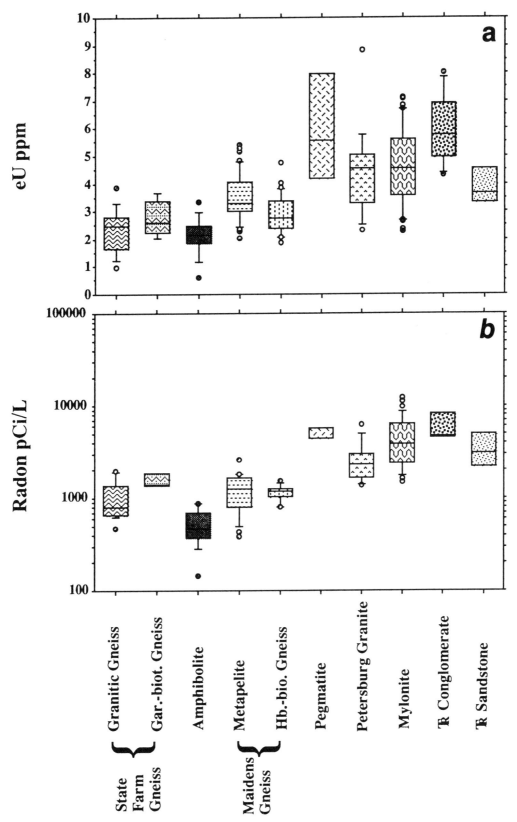

Figure 3. Graphs of rock types in the Hylas zone area as shown (see Table 1 for descriptions) vs. mean (line), 95 percent confidence standard deviation (box), and range of: a, equivalent uranium (eU, in parts per million); b, radon (in picoCuries per liter) in soils over the corresponding units.

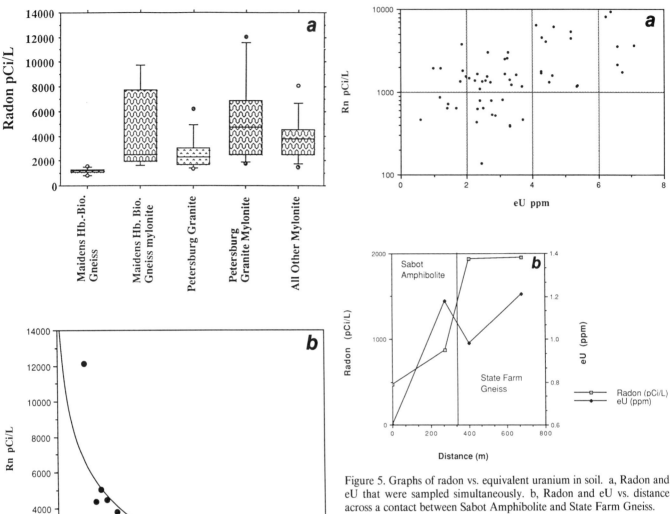

Figure 4. Graphs of soil radon over mylonites. a, Soil radon by rock type: deformed and undeformed. b, Soil radon vs. the angle between C and S bands in mylonitic Petersburg Granite.

Figure 5. Graphs of radon vs. equivalent uranium in soil. a, Radon and eU that were sampled simultaneously. b, Radon and eU vs. distance across a contact between Sabot Amphibolite and State Farm Gneiss.

mylonite was identified. Several other areas appear to require geologic reevaluation as a result of the radon survey. Another reason why soil radon surveys are reliable in the Hylas zone area is that the soils are saprolite derived directly from underlying bedrock. If soils are reworked by fluvial, coastal, eolian, or glacial processes, the relation between bedrock geology and both soil radon concentrations and gamma radiation is obscured (see Gates and others, 1990).

For soil radon and eU surveying to be effective, a contrast in uranium concentrations in the bedrock/soil units is required. Statistically, the degree of contrast may depend on the thickness of the feature to be identified, although a quantitative analysis of this relation has not been undertaken. The eU data is in general agreement with the geology in the Hylas area; however, the soil radon data appear to be a more accurate indicator of geologic patterns. Gamma-ray spectrometers measure radiation in the top 28 cm of soil (Durrance, 1986), depending on the radionuclide source, and is affected by mass of close objects such as outcrop. Further, redistribution of radionuclides by surface soil processes can give false high or low eU concentrations. Soil radon can be sampled at 75 cm or deeper, and in undisturbed soil and saprolite, is affected only by moisture. Both equivalent uranium and radon are affected by soils that have been disturbed by human activity. By choosing sampling sites carefully and conducting surveys during stable, dry weather, equivalent uranium and radon can add an important dimension to most geologic studies.

CONCLUSIONS

Both equivalent uranium and, especially, radon in soil directly reflect the chemistry of the underlying bedrock in areas where soil is derived from bedrock. Radon is a better indicator because it is sampled at a depth of 75 cm and is therefore less subject to surficial processes and human activity. Because it is relatively protected, soil radon appears to be a potentially powerful mapping tool in covered areas. Because of its dependence on small-scale geologic features, radon should not be evaluated using arbitrary political divisions, but rather the geology of an area.

ACKNOWLEDGMENTS

Support for this project was funded by the U.S. Department of Energy through the U.S. Geological Survey from Grant 05-88ER60665 (to L.C.S.G. and A.E.G.). Thanks to A. Attenborough, L. Hauser, G. Latske, L. Malizzi, and C. Wiggs for their many hours of field work. The suggestions of T. Baillieul, J. A. Speer, M. Stanton, and P. Whitney greatly enhanced the manuscript.

REFERENCES CITED

Baillieul, A., and Dexter, J. J., 1982, Evaluation of uranium anomalies in the Hylas zone and northern Richmond basin, east-central Virginia, *in* Goodknight, C. S., and Burger, J. A., eds., Reports on investigations of uranium anomalies: U.S. Department of Energy, National Uranium Resource Evaluation Report GJBX-222(82), 97 p.

Berthe, D., Choukroune, P., and Jegouzo, P., 1979, Orthogneiss, mylonite and non-coaxial deformation of granites: The example of the South American shear zone: Journal of Structural Geology, v. 1, p. 31–42.

Bobyarchick, A. R., 1976, Tectogenesis of the Hylas zone and eastern Piedmont near Richmond Virginia: Blacksburg, Virginia Polytechnic Institute and State University, (unpublished M.S. thesis), 168 p.

Bobyarchick, A. R., and Glover, L., III, 1979, Deformation and metamorphism in the Hylas zone and adjacent parts of the eastern Piedmont in Virginia: Geological Society of America Bulletin, v. 90, p. 739–752.

Bourland, W. C., 1976, Tectogenesis and metamorphism of the Piedmont from Columbia to Westview, Virginia, along the James River: Blacksburg, Virginia Polytechnic Institute and State University (unpublished M.S. thesis), 113 p.

Brown, C. B., 1937, Outline of the geology and mineral resources of Goochland County, Virginia: Virginia Division of Mineral Resources Bulletin, v. 48, 68 p.

Czarnecki, R. F., and Pacer, J. C., 1983, A comparative evaluation of radon measurement techniques for uranium exploration: U.S. Department of Energy Report GJBX-13(83), 63 p.

Durrance, E. M., 1986, Radioactivity in geology: Principles and applications: New York, Wiley & Sons, 441 p.

Farrar, S. S., 1984, The Goochland granulite terrane: Remobilized Grenville basement in the eastern Virginia Piedmont, *in* Bartholomew, M. J., ed., The Grenville Event in the Appalachians and related topics: Geological Society of America Special Paper, v. 194, p. 215–227.

Gates, A. E., and Glover, L., III, 1989, Alleghanian tectono-thermal evolution of the dextral transcurrent Hylas zone, Virginia Piedmont U.S.A.: Journal of Structural Geology, v. 11, p. 407–419.

Gates, A. E., and Gundersen, L.C.S., 1989, Role of ductile shearing in the concentration of radon in the Brookneal zone, Virginia: Geology, v. 17, p. 391–394.

Gates, A. E., Gundersen, L.C.S., and Malizzi, L. D., 1990, Comparison of radon in soil over faulted crystalline terranes: Glaciated versus unglaciated: Geophysical Research Letters, v. 17, p. 813–816.

Gingrich, J. E., and Fisher, J. C., 1976, Uranium exploration using the track etch method, *in* Proceedings: Exploration for Uranium Ore Deposits: Vienna, International Atomic Energy Agency, SM-208/19, p. 213–225.

Glover, L., III, Mose, D. G., Poland, F. B., Bobyarchick, A. R., and Bourland, W. C., 1978, Grenville basement in the eastern Piedmont of Virginia: Implications for orogenic models: Geological Society of America Abstracts with Programs, v. 10, p. 169.

Glover, L., III, Mose, D. G., Costain, J. K., Poland, F. B., and Reilly, J. M., 1982, Grenville basement in the eastern Piedmont of Virginia: A progress report: Geological Society of America Abstracts with Programs, v. 14, p. 20.

Grauch, R. I., and Zarinski, K., 1976, Generalized descriptions of uranium-bearing veins, pegmatites, and disseminations in nonsedimentary rocks, eastern United States: U.S. Geological Survey Open-File Report 76-582, 114 p.

Gundersen, L.C.S., Reimer, G. M., and Agard, S., 1988, Correlation between geology, radon in soil gas and indoor radon in the Reading Prong, *in* Marikos, M., ed., Proceedings of the GEORAD Conference: Geological Causes of Radionuclide Anomalies: Missouri Department of Natural Resources, p. 99–111.

Hand, B. M., and Banikowski, J. E., 1988, Radon in Onondaga County, New York: Paleohydrology and redistribution of uranium in Paleozoic sedimentary rocks: Geology, v. 16, p. 775–778.

Peacock, J. D., and Williamson, R., 1962, Radon detection as a prospecting technique: American Institute of Mining Engineers Transactions, v. 71, p. 75–85.

Platt, J. P., and Vissers, R.L.M., 1980, Extensional structures in anisotropic rocks: Journal of Structural Geology, v. 2, p. 397–410.

Poland, F. B., 1976, Geology of the rocks along the James River between Sabot and Cedar Point, Virginia: Blacksburg, Virginia Polytechnic Institute and State University (unpublished M.S. thesis), 98 p.

Ramsay, J. G., and Graham, R. H., 1970, Strain variation in shear belts: Canadian Journal of Earth Science, v. 7, p. 786–813.

Reilly, J. M., 1980, A geologic and potential field investigation in the central Virginia Piedmont: Blacksburg, Virginia Polytechnic Institute and State University (unpublished M.S. thesis), 111 p.

Reimer, G. M., 1992, Simple techniques for soil-gas and water sampling for radon analysis, *in* Gundersen, L.C.S., and Wanty, R. B., eds., Field studies of radon in rocks soils and water: U.S. Geological Survey Bulletin 1971.

Severne, B. C., 1978, Evaluation of radon systems at Yeelirie, Australia: Journal of Geochemical Exploration, v. 9, p. 1–22.

Simpson, C., and Schmid, S. M., 1983, An evaluation of criteria to deduce the sense of movement in sheared rocks: Geological Society of America Bulletin, v. 94, p. 1281–1288.

Smith, A. Y., Barretto, P.M.C., and Pournis, S., 1976, Radon methods in uranium exploration, *in* Proceedings: Exploration for Uranium Ore Deposits: Vienna, International Atomic Energy Agency, SM-208/52, p. 185–211.

Weems, R. E., 1974, Geology of the Hanover Academy and Ashland quadrangles, Virginia: Blacksburg, Virginia Polytechnic Institute and State University (unpublished M.S. thesis), 98 p.

Wright, J. E., Sinha, A. K., and Glover, L., III, 1975, Age of zircons from the Petersburg Granite, Virginia: With comments on belts of plutons in the Piedmont: American Journal of Science, v. 275, p. 848–856.

Manuscript Accepted by the Society April 6, 1992

Geologic and environmental implications of high soil-gas radon concentrations in the Great Valley, Jefferson and Berkeley counties, West Virginia

Art Schultz
U.S. Geological Survey, 926 National Center, Reston, Virginia 22092
Calvin R. Wiggs
HydroGeoLogic Inc., 503 Carlisle Drive, Suite 250, Herndon, Virginia 22070
Stephen D. Brower
Department of Geology, West Virginia University, Morgantown, West Virginia 26505

ABSTRACT

Soil-gas radon and ground radioactivity surveys across a portion of the Great Valley of West Virginia indicate that residuum and soils formed above some carbonate rocks have sufficient levels of radon gas to cause high indoor radon values. Data indicate no correlation of soil-gas radon concentration with faults, cleavage, joints, or calcite veins. Instead, soil-gas radon distribution appears to be controlled by the solution of carbonate bedrock and the subsequent development of thick, red, clay-rich residuum, which may contain as much as 4 times the concentration of radium, 10 times the concentration of uranium, and 5 times the concentration of thorium as the underlying bedrock. Such residuum and associated soil develops over some parts of the Elbrook, Conococheague, and Beekmantown Formations, and can have concentrations of radon in soil-gas exceeding 4,000 pCi/L. In areas of the Great Valley underlain by siltstone, fine-grained sandstone, and shale of the Martinsburg Formation, soil-gas radon values can exceed 4,000 pCi/L. In these areas, bedrock alone appears to have sufficient thorium, radium and uranium concentrations to generate the soil-gas radon measured. Previous work by others and our own preliminary evaluations indicate that soil-gas radon levels are high enough to cause indoor air in homes to exceed 4 pCi/L, the U.S. Environmental Protection Agency's (EPA) action level for radon. Aeroradiometric maps and National Uranium Resource Evaluation (NURE) Program data do indicate anomalously high radioactivity in some areas where radon soil-gas concentrations were high. These data, used with available geologic maps, soil maps, and maps showing thickness of residuum, are useful in predicting areas of radon soil-gas hazards.

INTRODUCTION

Limestone and dolomite, like that which underlies much of the Great Valley of West Virginia, generally does not contain significant amounts of uranium (Gabelman, 1977). However, over the past 5 yr, the U.S. Environmental Protection Agency (EPA) and local health officials have shown that homes in the Great Valley of West Virginia, an area known to be underlain by carbonate bedrock, have levels of indoor radon that exceed government guidelines. This chapter examines the concentrations of radon in soil-gas across the Great Valley and determines the relation of soil-gas radon to mapped geologic formations, structures, and surficial materials. A predictive model relating the distribution of radon in soil-gas to the underlying bedrock and surficial geologic materials is presented. This model can be used to evaluate potential radon hazards across the Great Valley in

Schultz, A., Wiggs, C. R., and Brower, S. D., 1992, Geologic and environmental implications of high soil-gas radon concentrations in the Great Valley, Jefferson and Berkeley counties, West Virginia, *in* Gates, A. E., and Gundersen, L.C.S., eds., Geologic Controls on Radon: Boulder, Colorado, Geological Society of America Special Paper 271.

West Virginia and in similar terranes elsewhere. Finally, the chapter considers existing data bases, such as aeroradioactivity surveys, soil surveys, and geologic maps, for their effectiveness in predicting regional scale radon distribution.

PHYSIOGRAPHIC AND GEOLOGIC SETTING

The Great Valley of West Virginia is bordered on the east by the Blue Ridge Mountains and on the west by the mountains of the Appalachian Valley and Ridge Province (Fig. 1). Most of the Great Valley in the study area is characterized by gently rolling topography with low, parallel, linear ridges, trending northeast-southwest, that reflect the underlying bedrock geology. Although the area is rural, rapid residential and industrial development is occurring and is expected to increase in the near future.

Bedrock underlying the Great Valley consists of 14,000 ft (425 m) of carbonate and siliciclastic rocks ranging from Lower Cambrian to lower Middle Ordovician age (Fig. 2). The section is composed of dolomite and limestone with subordinate amounts of cherty limestone and dolomite, shaly and silty dolomite and limestone, calcareous siltstone, calcareous fine-grained sandstone and siliceous fine-grained sandstone, siltstone, and shale. Thermal maturation ranges from 250° to 300° (Epstein and others, 1977).

The dominant regional structure in the Great Valley of the study area is the Massanutten synclinorium (Figs. 1, 3). This large-scale structure and the associated deformation are the result of northwest-directed thrusting during the late Paleozoic Alleghenian orogeny. The synclinorium is bounded on the northwest by the North Mountain fault, a regional scale southeast-dipping thrust that can be traced from southern Pennsylvania into

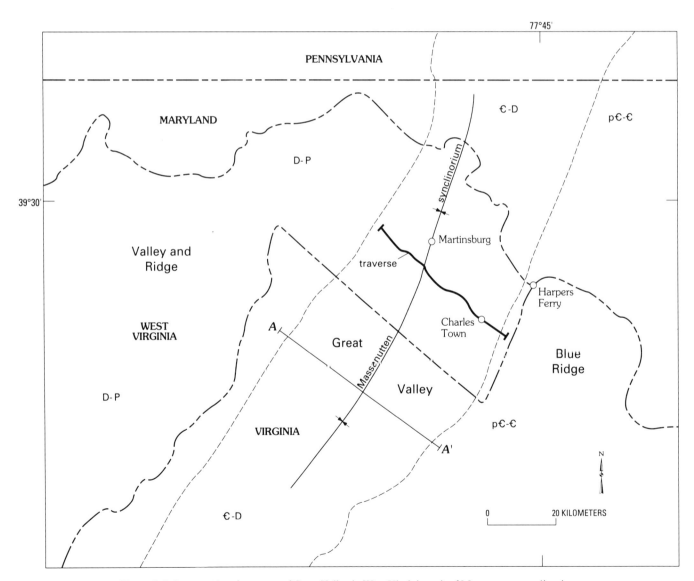

Figure 1. Index map showing extent of Great Valley in West Virginia, axis of Massanutten synclinorium, position of soil-gas radon traverse (Figs. 4, 5, 8) and regional scale cross section A–A' (Fig. 3).

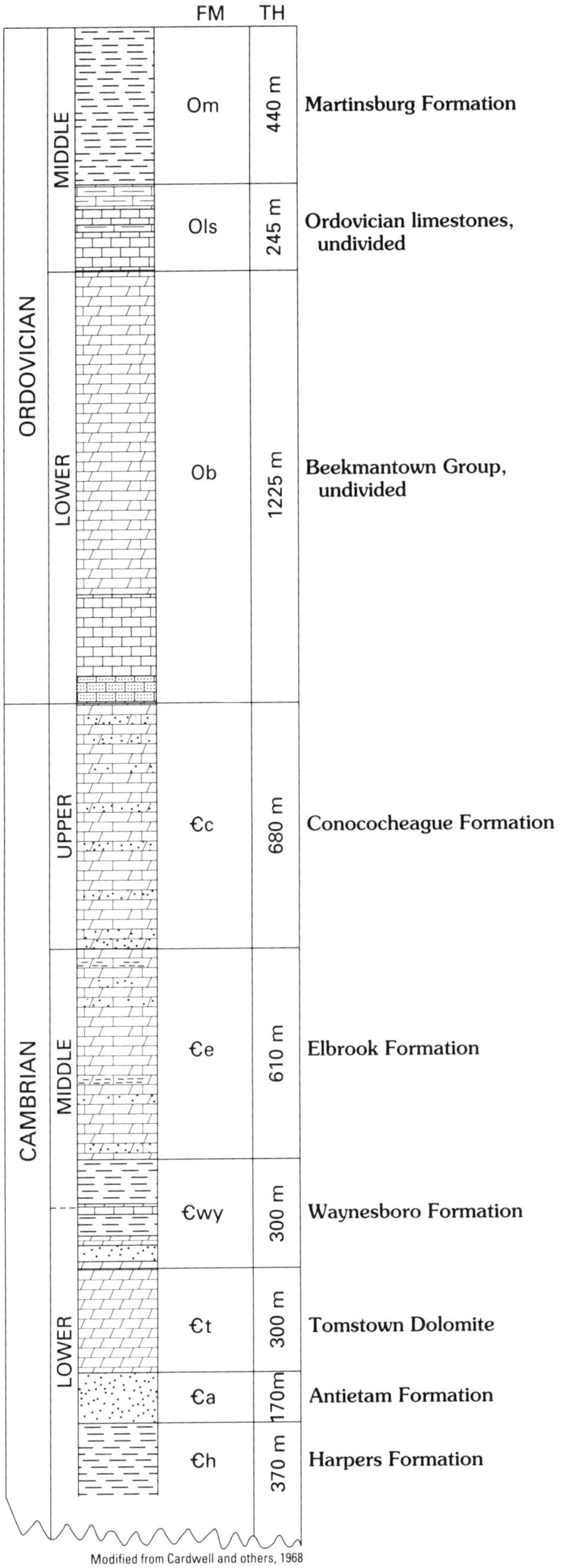

southwest Virginia. The fault places Lower Cambrian dolomite on Ordovician, Silurian, and Devonian rocks of the Valley and Ridge Province.

In the study area, the southeast limb of the synclinorium is thickened by folding and faulting (Fig. 4). All formations are repeated across the axis of the fold except the Waynesboro Formation and Tomstown dolomite, which are truncated by the North Mountain fault in the subsurface. Folds and faults are present in the study area and range in size from mesoscopic to megascopic. In general, the folding is disharmonic and asymmetric, with northwest vergence in the direction of major tectonic transport. Many of the folds within the study area are associated with thrust faults (Figs. 3, 4). These faults dip consistently to the southeast. Numerous cross-strike faults of minimal displacement occur across the study area (Fig. 4).

Pressure solution cleavage is found throughout the area but is well developed in the shaly carbonates and calcareous shales. The most intense cleavage occurs near faults.

Joints occur in all rock types in the Great Valley. One conspicuous joint set trends northwest-southeast and is vertical to steeply dipping. These joints are approximately perpendicular to the regional strike of bedrock.

Surficial geologic materials in the Great Valley include alluvial deposits, terrace deposits, residuum, and minor colluvium. Alluvial, terrace, and colluvial deposits consist of cobbles, pebbles, sand, silt, and clay. These materials are not extensive in the study area and were not sampled in this study.

Residuum is extensive across the Great Valley and is highly variable in thickness (Fig. 5). Three general types of residuum and associated soils have been recognized in the Great Valley of the study area (Hack, 1965; Gorman and others, 1966; Hatfield and Warner, 1973). Type 1 residuum overlies Ordovician shale, siltstone, and very fine-grained sandstone of the Martinsburg Formation and colluvium and alluvium along major drainages and below mountain slopes. This residuum is a thin, patchy mantle with bedrock generally less than a meter from the surface. The residuum consists of silt and clay grading downward into loose fragments of bedrock. Soils in this area are dominantly shallow and poorly drained, and have been derived from weathering of the underlying siliciclastic material.

Type 2 residuum overlies limestone of Middle and Lower Ordovician age. This residuum is thin and patchy and is interrupted by many outcrops at the surface (Fig. 6). The residuum varies in thickness from a few centimeters to more than a meter and is generally a silty loam with mixed fragments of limestone and minor chert. The soils are variable in thickness, and outcrops may be extensive.

Type 3 residuum overlies dolomite and limestone rock of Cambrian age that have a greater amount of argillaceous and

Figure 2. Lithologic and stratigraphic column (modified from Cardwell and others, 1968) of rocks in the Great Valley in study area.

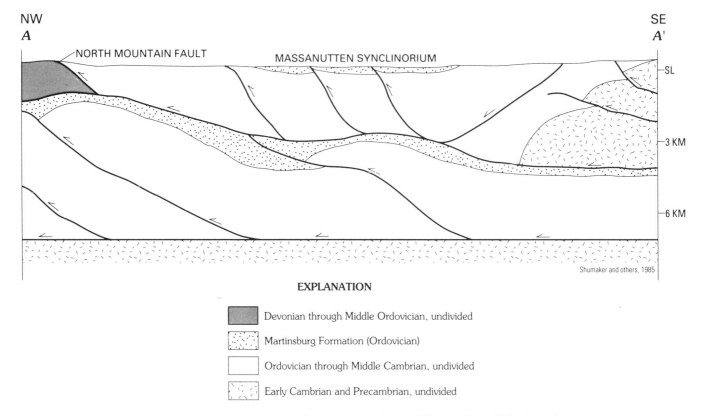

Figure 3. Regional scale cross section (Shumaker and others, 1985) of the Great Valley in study area vicinity, showing the allochthonous Massanutten synclinorium. Location of section in Figure 1.

siliciclastic beds than does the overlying Ordovician limestone. This residuum is thick and in places may exceed 20 m (Hack, 1965). Bedrock outcrops in the areas of thick residuum are limited (Fig. 7) and consist of pinnicles. Sandy, clayey, and silty loams are characteristic, and a persistent residual concentrate of chert and cherty sandstone fragments, as thick as a meter, occurs on ridges throughout this part of the Great Valley. Soils are usually deep and well drained.

METHODS

To characterize the distribution of radon in soil-gas across the Great Valley of West Virginia, a traverse approximately perpendicular to the regional strike of bedrock was chosen (Figs. 1, 8). The northwest-southeast traverse crosses parts of the Lower Cambrian through Middle Ordovician sequence.

One hundred and forty-four soil-gas radon measurements were made at 1-km spacings along a traverse near West Virginia Routes 51 and 9. Sample sites were at least 10 m from the roads and disturbed ground was avoided. Soil-gas radon was measured by inserting a steel probe into the soil to a depth of 75 cm and withdrawing 20 cc of gas. Samples were then measured in an alpha-sensitive scintillometer adapted with a Lucas cell and counted for 2 min (see Reimer, 1992, for a complete description of this sampling method). Samples were measured in July during a 5-day period of warm, dry weather under generally stable high barometric pressure following 3 days of no rain. Total count radioactivity readings were recorded for the surface of the soil or residuum and on rock outcrops with a hand-held scintillometer. Several measurements at 3-min counting intervals were made at each location and averaged.

Eleven bedrock and 23 soil samples were collected at several soil-gas radon sample localities along the traverse in each of the stratigraphic units for analysis of uranium, potassium, radium, and thorium. Bedrock samples were taken from the surface and broken from the outcrop with a rock hammer. At each sample location, soil was taken with an 8-cm-diameter hand auger. Each sample was 10 cm long and taken from depths of 10 to 20 cm below the surface and from 165 to 175 cm below the surface. Chemical uranium was determined by fluorimetry and radium by deemanation. Thorium and potassium were determined by spectroscopy. Analyses were performed by Natural Resources Lab, Golden, CO.

DATA

Total count radioactivity (Fig. 8) along the entire traverse and for both bedrock and soil ranged from <10 to about 40 cps. In general, bedrock was low in total radioactivity and consistent

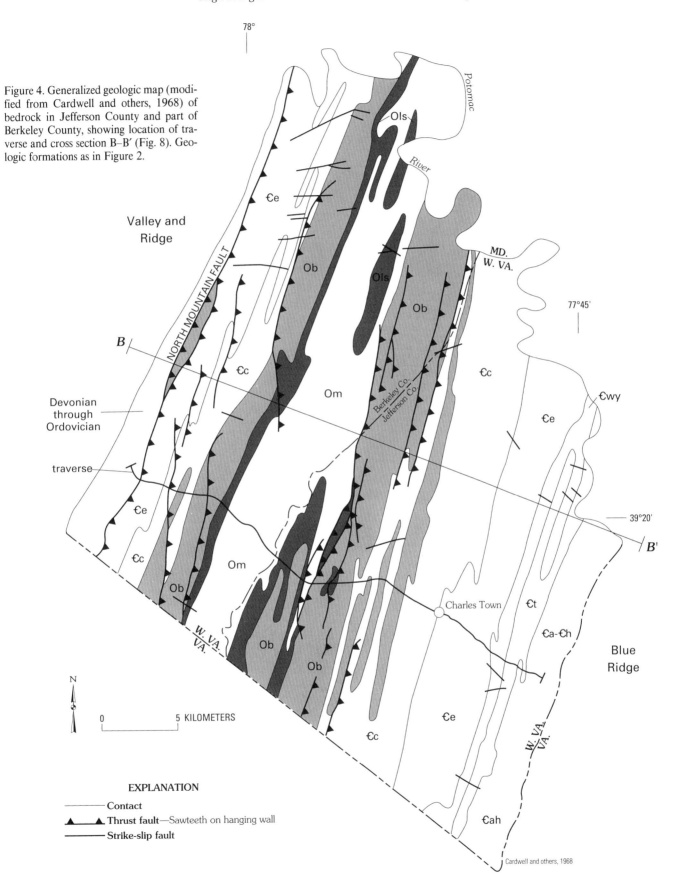

Figure 4. Generalized geologic map (modified from Cardwell and others, 1968) of bedrock in Jefferson County and part of Berkeley County, showing location of traverse and cross section B–B′ (Fig. 8). Geologic formations as in Figure 2.

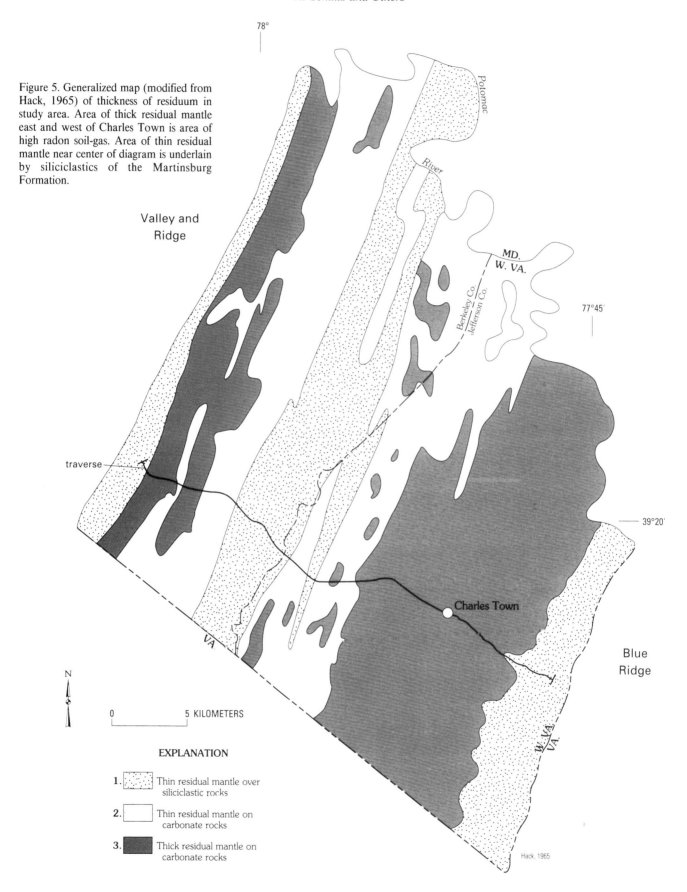

Figure 5. Generalized map (modified from Hack, 1965) of thickness of residuum in study area. Area of thick residual mantle east and west of Charles Town is area of high radon soil-gas. Area of thin residual mantle near center of diagram is underlain by siliciclastics of the Martinsburg Formation.

Figure 6. Photograph of typical topography in area of thin residuum over carbonate bedrock. Numerous bedrock pinnacles protrude from the thin residual deposits.

Figure 7. Photograph of area with thick residuum, gently rolling topography, and few outcrops on the surface. Linear ridge in background, in front of silo, is held up by silty and cherty carbonate rocks. Ridge in distant background is Blue Ridge Mountains.

with carbonates elsewhere. The data (Fig. 8) show that total count radioactivity for carbonate bedrock was consistently lower than that for the surrounding soil and residuum. Total count on shale outcrops was consistently higher than the surrounding and overlying soil and residuum. Also, outcrops of argillaceous carbonates and carbonates with thin silty layers were generally higher in total count radioactivity than relatively pure limestone and dolomite with little siliciclastic component.

Total count radioactivity on fractured, veined, and faulted rocks along the traverse were similar to the nearby undeformed rocks. One fault along the traverse is well exposed. The fault places older Middle Ordovician limestone over younger Middle Ordovician limestone. Cleavage, fractures, and calcite veins are concentrated in both the hanging wall and footwall within 20 m of the fault surface. Continuous scintillometer readings across the fault zone showed no increase in total radioactivity. Similar results were found across several mapped faults on the traverse.

Soil-gas radon concentrations (Fig. 8, Table 1) ranged from

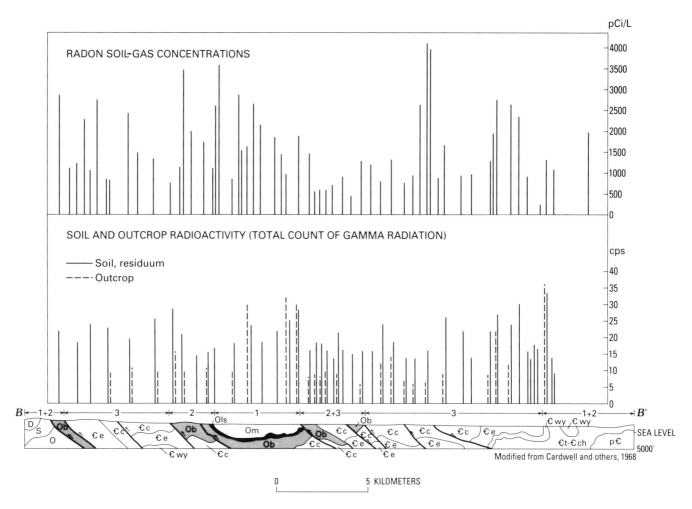

Figure 8. Total count gamma-ray radiation data and radon soil-gas concentrations along the traverse. Approximate location of data collection site is plotted on cross section B–B' near the traverse. Location of cross section B–B' is shown in Figure 4. Total counts in carbonate bedrock are consistently lower than overlying residuum. Total counts in siliciclastic rocks are consistently higher than overlying residuum.

about 500 pCi/L to more than 4,000 pCi/L for the study area. In general, the lowest concentrations were in areas underlain by limestone of Middle Ordovician age. The highest concentrations were in areas underlain by shaly dolomite and shaly, silty limestone and dolomite of Lower Ordovician and Cambrian age and in areas underlain by shale, siltstone, and fine-grained sandstone of Middle Ordovician age.

Soil-gas radon values exceeding 2,500 pCi/L occurred in a 2-km-wide zone along the traverse just to the south of Charles Town. The highest values occurred in soils on ridges underlain by argillaceous, silty, and cherty dolomite.

Soil-gas radon concentrations between 2,000 and 4,000 pCi/L were found near the axis of the Massanutten synclinorium in soils underlain by shale, siltstone, and fine-grained sandstone of the Ordovician Martinsburg Formation.

High soil-gas radon concentrations between 3,500 and 4,000 pCi/L occurred in soils underlain by rocks of the lower and middle parts of the Conococheague Formation in interbedded limestone, cherty dolomite, and silty dolomite. Detailed site work at this location outlined the distribution of the anomaly (Fig. 9). The area of high soil-gas is about 100 m wide and 500 m long as outlined by our sampling.

Soil-gas radon concentrations across mapped and observed faults (Fig. 10) showed no systematic change. Values were consistent with surrounding rocks where no faults have been mapped.

Geochemical analysis of samples (Tables 2, 3) show that soil and residuum above carbonate rocks contain more uranium, potassium, radium, and thorium than the adjacent and underlying bedrock. Soils above shales and siltstones have similar concentra-

TABLE 1. RADON SOIL-GAS VALUES FROM RESIDUUM OVERLYING SELECTED BEDROCK FORMATIONS

Stratigraphic Unit	Radon Soil-Gas Values (pCi/L)
Tomstown Dolomite	2,220, 2,065, 1,783, 2, 527, 1,105
Waynesboro Formation	904, 1,020, 1,164, 1,433, 1,641, 266, 1,331
Elbrook Formation	1,536, 3,206, 2,795, 1,766, 804, 2,044, 1,959, 1,211, 1,803, 899, 1,541, 832, 1,355, 1,059, 1,261, 961, 2,359, 2,669, 2,720, 1,295, 875, 1,156
Conococheague Formation	1,063, 1,379, 1,778, 787, 1,487, 1,291, 744, 1,643, 1,254, 1,697, 2,097, 2,175, 872, 1,685, 367, 2,779, 4,185, 1,160, 997, 2,822, 1,008, 901, 558, 971, 939, 1,628, 4,094, 939, 758, 740, 3,440, 2,260, 1,612, 1,950, 2,040, 1,160, 490, 1,291, 778, 1,175, 1,260, 423, 1,178, 2,331, 593, 3,156, 1,719
Beekmantown Group	1,741, 2,179, 932, 2,246, 1,025, 82, 2,116, 1,521, 741, 1,082, 516, 1,479, 1,976, 2,657, 1,691, 3,764, 720, 879, 694, 602, 578, 1,449, 2,121, 564, 1,195, 1,001, 708, 2,701
Middle Ordovician limestone, undifferentiated	873, 788, 1,680, 506, 435, 1,252
Martinsburg Formation	288, 4,004, 3,090, 3,094, 721, 928, 2,255, 508, 395, 2,316, 535, 417, 2,229, 2,542, 843, 1,657, 199, 1,466, 650, 1,836, 3,887, 507, 933, 1,835, 1,707, 2,123, 1,617, 852, 1,098

tions of uranium, potassium, and thorium as the adjacent bedrock. Within our sample population, uranium appears depleted and in disequilibrium with radium-226 in both the bedrock and soil or residuum (Figs. 11, 12). Uranium is also somewhat depleted from the surface samples of soil or residuum when compared with samples from the same location at depth (Fig. 13).

Airborne total count gamma-ray radioactivity data (Neuschel, 1965) (Fig. 14) for part of the traverse correlate in some areas to ground radioactivity and soil-gas radon concentration. For example, some of the highest radioactivity on the airborne traverse (700 to 800 cps) occurs just south of Charles Town (Fig. 14). This area has soil-gas radon concentrations in excess of 2,500 pCi/L. Conversely, a small but high soil-gas radon concentration in the lower and middle part of the Conococheague Formation does not show up on the same airborne data base. This area does appear to correlate closely to available airborne uranium spectral data (Texas Instruments, Inc., 1978) where a uranium peak is located in the vicinity of the highest soil-gas radon concentrations (Fig. 15).

INTERPRETATION

Within the study area, soil-gas radon distribution appears to be controlled by the underlying bedrock and the weathered material derived from the bedrock. We found no evidence of unusually high radioactivity in bedrock or along joints or faults. Leaching of bedrock uranium may be inferred from our geochemical data (Figs. 11, 12) that show depletion of uranium with respect to Ra-226. Alternatively, enrichment of Ra-226 relative to uranium has taken place. More sampling is needed to determine if the depletion is only in the upper meter of exposed outcrops (usually the portion that we sampled) or if disequilibrium has occurred throughout the carbonates of the Great Valley. Regional scale disequilibrium between uranium and radium may have occurred during mass movement of fluids associated with the Alleghenian orogeny. Such regional scale mass movement of fluids has been documented for carbonate rocks in the Great Valley (Hearn and others, 1987).

Our geochemical analysis of carbonate bedrock, although limited in scope, shows that carbonate rocks with the greater percentage of detrital material (i.e., clay, silt, and sand) have higher radioactivities and higher concentrations of uranium than carbonates with little or no siliciclastic component. Regionally, shale and siltstone have the highest total count radioactivity and have greater amounts of potassium, uranium, radium, and thorium than carbonates. We assume that the higher amounts of these elements in both carbonate and noncarbonate siliciclastics are chiefly due to the higher concentrations of detrital clay minerals and to higher amounts of detrital uranium-bearing minerals and primary uranium derived from reducing depositional environments. Our data on cleaved, fractured, faulted, and calcite mineralized rocks in the study area indicate no increase in uranium content associated with these features. Perhaps uranium was not mobilized or concentrated during or after faulting or fracturing, or perhaps uranium emplaced in the fractures and faults has subsequently been removed. These results are substantially different than those obtained in recent studies of mylonite zones developed in metamorphic rocks (Gates and Gundersen, this volume). In mylonitization, substantial amounts of recrystallization and chemical changes occur, which concentrate uranium. Faults of the Great Valley are characteristically cataclastic zones of brittle deformation with little recrystallization; i.e., little change in the bulk chemistry of the rock occurs during deformation. Calcite, which fills most of the fractures near fault zones in the Great Valley, is probably derived from pressure solution of nearby carbonates. Our data indicate no increase or decrease in uranium content associated with this transfer of calcite.

Total count radioactivity of soils and residuum across the study area reflects both the type of bedrock and character of

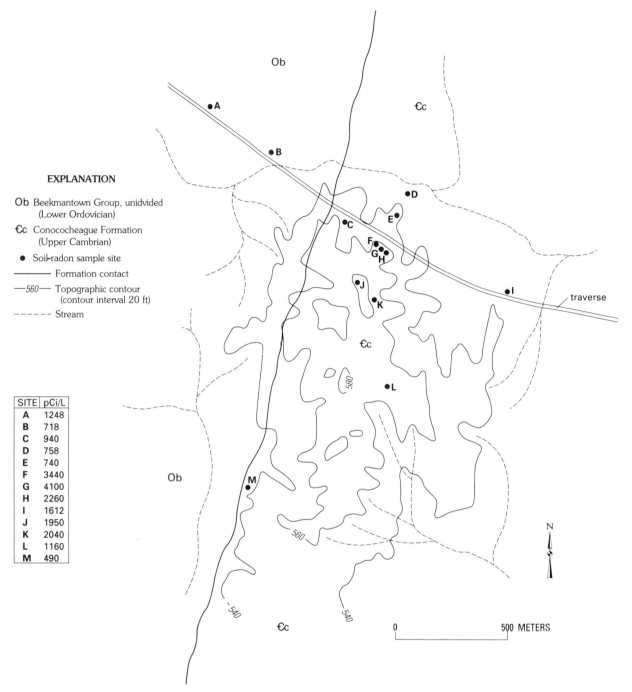

Figure 9. Map outlining an area of high soil-gas radon concentration. The highest concentrations occur away from the drainages and on a flat topographic area near the top of the linear ridge. Note the topography is parallel to the geologic contact. The ridge is covered with a mantle of chert and silty carbonate clasts.

residual material derived from weathering of the bedrock. Weathering in the carbonate rocks consists of dissolution of calcium and magnesium carbonate with subsequent concentration of residual materials to form the overlying residuum and soil. The thickness of this residuum and soil is highly variable across the carbonate rocks; however, the concentration of uranium, potassium, radium, and thorium in the residuum is consistently higher than the underlying bedrock in the majority of the sample localities. These results are consistent with detailed site analysis of carbonate rocks in the Great Valley elsewhere (Greenman and

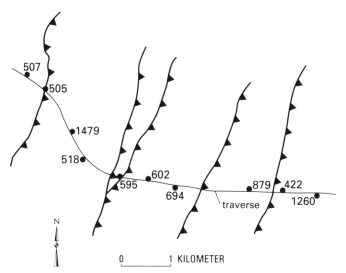

Figure 10. Soil-gas radon concentrations plotted along a portion of the traverse that crosses several mapped faults. No apparent increase in radon soil-gas concentration is associated with the faults.

TABLE 2. SELECTED GEOCHEMICAL ANALYSES OF ROCKS FROM THE GREAT VALLEY OF WEST VIRGINIA

Formation / Rock Type	U (ppm)	Th (ppm)	K (%)	Ra-226 (pCi/g)
Tomstown Dolomite				
Dolomite	0.6	3	0.8	0.5
Waynesboro shale				
Shale	1.5	10	7.9	1.0
Elbrook Formation				
Limestone	0.7	<3	1.8	0.3
Shale	1.7	6	5.4	0.7
Limestone / Dolomite	1.8	4	3.6	0.3
Conococheague Formation				
Limestone	1.6	<3	0.6	0.6
Limestone	0.3	<3	<0.2	0.3
Beekmantown Group				
Limestone / Dolomite	0.6	<3	0.2	0.3
Middle Ordovician Limestone, undifferentiated				
Limestone	4.5	<3	<0.2	2.2
Limestone	0.3	<3	<0.2	0.3
Martinsburg Formation				
Shale	2.8	14	3.5	1.3

Note: Limits of detection are shown below: Uncertainties (1 standard deviation) equal limit of detection or 10% of the reported value, whichever is larger.
U = 0.1 ppm; Th = 3.0 ppm; K = 0.2%; Ra-226 - 0.3 pCi/g.

TABLE 3. GEOCHEMICAL ANALYSES OF SOIL AND RESIDUUM OVERLYING SELECTED ROCK TYPES FROM THE GREAT VALLEY OF WEST VIRGINIA

Formation and Rock Type	Soil / Residuum	U (ppm)	Th (ppm)	K (%)	Ra-226 (PCi/g)
Tomstown Dolomite					
Dolomite	Surface	3.2	9	1.9	1.0
	0.75 m down	2.5	10	1.2	1.0
Waynesboro Formation					
Shale	Surface	1.6	13	6.4	0.8
Elbrook Formation					
Dolomite	Surface	1.5	9	3.4	0.7
	0.75 m down	2.3	14	1.6	0.8
Dolomite	Surface	2.8	18	5.2	1.2
	0.75 m down	3.4	15.8	4.8	1.3
Shale	Surface	2.8	10	5.8	1.2
	0.75 m down	3.1	17	5.8	0.5
Limestone / Dolomite	Surface	3.7	11	6.7	1.2
	0.75 m down	3.2	11	7.3	1.2
Conococheague Formation					
Limestone	Surface	2.1	10	5.3	0.8
	0.75 m down	2.5	10	4.1	0.9
Limestone	Surface	3.2	12	1.8	0.7
	0.75 m down	3.1	16	1.9	0.9
Beekmantown Group					
Limestone / Dolomite	Surface	1.1	3	1.7	1.0
	0.75 m down	1.6	10	2.3	1.3
Middle Ordovician Limestone, undifferentiated					
Limestone	Surface	1.9	16	1.8	0.5
	0.75 m down	1.5	12	1.9	1.0
Martinsburg Formation					
Shale	Surface	2.3	13	2.6	1.0
	0.75 m down	2.2	11	3.1	0.9
Shale	Surface	1.9	10	2.9	1.3
	0.75 m down	2.9	11	3.4	1.4

Note: Limits of detection are shown below: Uncertainties (1 standard deviation) equal limit of detection or 10% of the reported value, whichever is larger.
U = 0.1 ppm; Th = 3 ppm; K = 0.2%; Ra-226 = 0.3 pCi/g.

others, 1990), which showed a 10-fold increase of uranium in soils over bedrock. Because shaly, silty, and sandy carbonate rocks have a greater amount of detrital mica, uranium-bearing minerals, and uranium adsorbed onto clays and organics, the soils and residuum formed from their solution are richer in uranium, thorium, and potassium than the solution residue of carbonate rocks with lesser amounts of primary detrital constituents.

In the Great Valley, in areas underlain by shale, siltstone, and fine-grained sandstone, weathering consists chiefly of mechanical and chemical breakdown along the outcrop surface. Residuum consists of fragments of bedrock mixed with quartz silt and sand, and clay. Where topographic relief is considerable, residuum is thin and patchy and is easily washed from the slopes. Little of the residual material is preserved above the bedrock; thus concentration of uranium, potassium, radium, and thorium in the soils is similar to underlying bedrock. Thicker residual deposits

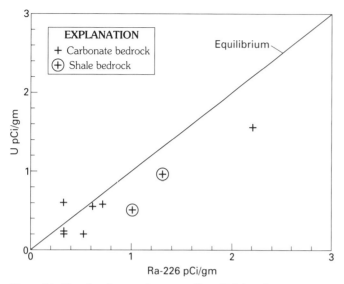

Figure 11. Plot showing uranium vs. radium-226 data from carbonate and shale along the traverse. Uranium is depleted with respect to radium.

and soils may form in areas underlain by siliciclastic rocks in areas of low relief. Data indicate that this material is not enriched in uranium with respect to the underlying bedrock.

One area of high soil-gas radon concentration (Fig. 9) is underlain by interbedded limestone and dolomite with thin silty, cherty beds in the upper part of the Cambrian Conococheague Formation. Data in this area indicate a probable correlation between the thickness and type of residuum and soil and this anomaly. The highest concentrations were obtained on a topographic bench near the top of a discontinuous linear ridge. Residuum exposed in this area by recent construction is 5 m thick. Data along strike, both to the northeast and southwest of the area of high soil-gas radon, showed a rapid decrease of soil gas-radon concentration. Residuum is thickest in the area of the high soil-gas radon concentration and is thinner away from this anomaly. Furthermore, although bedrock is sparse in this area, total count radioactivity on outcrops along strike of the soil-gas anomaly showed no change. We assume no increase in bedrock concentration of uranium along strike. Bedrock within a meter of soil-gas radon concentrations of 4,200 pCi/L is <10 cps total count radioactivity. Several other outcrops within the zone of highest radon soil-gas were also <20 cps.

The high soil-gas radon concentrations (>4,000 pCi/L) were in those areas of chert and silt-bearing limestone and dolomite where red, clay-rich residuum was thick. Soil-gas radon concentrations along strike changed as the thickness and character of residuum changed. Thin, yellow and brown silty soils along strike showed low radon soil-gas values. Similarly, thin, red, clay-rich residuum along strike also had lower soil-gas radon concentrations.

The preceding observations lead us to suggest that several factors probably relate to the distribution of radon soil-gas.

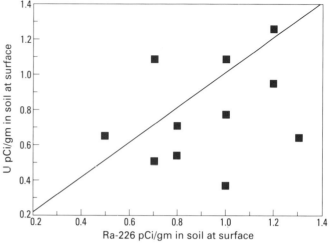

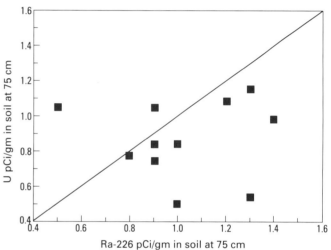

Figure 12. Plots showing uranium vs. radium-226 data from soil and residuum near the surface and at 75 cm below the surface. Note the trend of uranium depletion in both near-surface and at depth.

Higher concentrations of radon in soil-gas is expected where the soil or residuum is thicker, more red, and more clay-rich. Across the Great Valley, such soil conditions are repeatedly observed on the linear topographic ridges located over carbonate rocks of the Elbrook, Conococheague, and Beekmantown Formations.

The formation of thick, red, clay-rich residuum in the Great Valley carbonates is controlled by interbedded siliceous material such as clay, silt, fine-grained sand, and chert (Fig. 16). During dissolution of the carbonate bedrock, siliceous chert and calcareous siltstone and sandstone accumulated on the surface and formed a layer armoring and preventing the finer grained residuum (derived from bedrock solution) from being washed into the adjacent valleys. In this way, deep residual deposits accumulated and formed resistant ridges. Within the carbonate rocks of the Great Valley, silty, cherty layers commonly form discontinuous lenses. This model suggests one possible reason for the un-

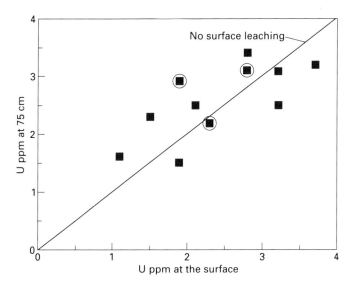

Figure 13. Plot showing amounts of uranium in soil and residuum at near-surface and 75-cm depths. Circled data are from soils underlain by shales, uncircled are from carbonates. Uranium is slightly leached in the surface samples.

even distribution of thick residuum and the variability of radon soil-gas concentrations along strike.

Water and indoor radon surveys of houses along the traverse are presently underway. Radon in water up to 2,000 pCi/L was measured in a sampling of 97 domestic water wells (Brower, 1991). The majority of concentrations were below 1,000 pCi/L. The highest water radon concentrations were measured in wells sited in siltstones and shales interbedded with the carbonate rocks. Most wells are drilled into bedrock in the Great Valley. Since the carbonate bedrock is low in uranium concentration, we expect little input from bedrock to circulating ground water. Preliminary indoor radon values show that approximately 25 percent of the homes tested exceed the EPA's action level of 4 pCi/L (Brower, 1991). Many of these homes are found in areas underlain by residuum with high radon soil-gas anomalies.

ENVIRONMENTAL IMPLICATIONS

Empirical data from studies of basements in homes in the northeast U.S. (Gundersen, 1989, personal communication, 1990) show that indoor radon concentrations are generally 0.6 to

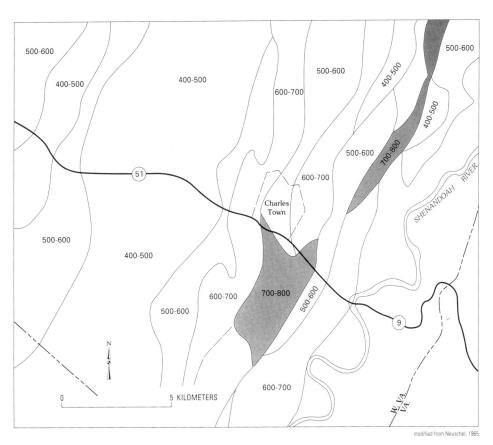

Figure 14. Regional scale aeroradioactivity map (modified from Neuschel, 1965) for part of the study area. Relatively high area just south of Charles Town (700 to 800 cps) was found to have relatively high radon soil-gas concentrations.

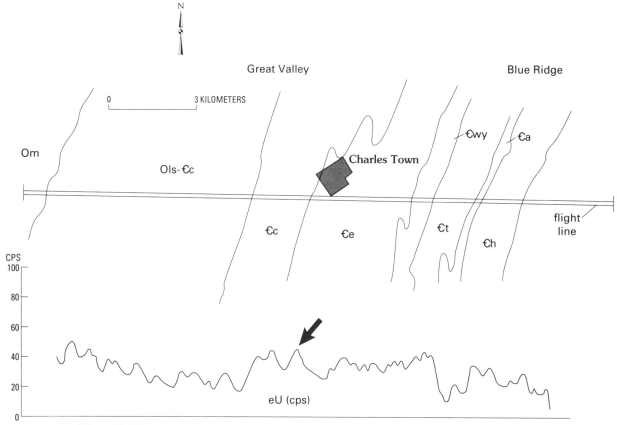

Figure 15. Airborne uranium data from the National Uranium Resource Evaluation Program (NURE) (Texas Instruments, Inc., 1978) for a portion of the traverse. Arrow indicates peak in equivalent uranium concentration, which correlates closely with high concentrations of soil-gas radon found on the traverse.

10 percent of the radon concentration measured in the soil with an average of 1 percent. Radon soil-gas concentration above 500 pCi/L may cause indoor radon levels exceeding the EPA recommended action level of 4 pCi/L. Levels higher than 2,000 pCi/L of radon soil-gas may cause notably high levels of indoor radon. Several locations on the traverse have soil-gas radon concentrations that exceed 2,000 pCi/L. These were found in the upper part of the Elbrook Formation, in the lower and middle part of the Conococheague Formation, in part of the Beekmantown Formation, and in part of the Martinsburg Formation.

The results of this study indicate that indoor radon problems may occur if houses are built on thick, clay-rich, red residuum that has developed over argillaceous, cherty, and silty limestone and dolomite of the Elbrook, Conococheague, and Beekmantown Formations. Where houses are built directly on carbonate rocks with thin or nonexistent residual deposits, we expect little radon input from bedrock and consequently little probability of indoor radon problems. In addition, homes in areas underlain by siltstone, fine-grained sandstone, and shale of the Martinsburg Formation may have radon concentrations exceeding the EPA action level. In these areas, construction on bedrock and on the soil or residuum derived from the bedrock pose similar problems. This is so, since bedrock may contain the same or higher amounts of uranium than weathered bedrock and residuum or soil.

Ground water results (Brower, 1991) indicate that wells in the Great Valley probably do not have radon problems. Results of indoor radon testing along the traverse (Brower, 1991) indicate indoor values that exceed the EPA action level occurred in about 25 percent of the homes.

SUMMARY AND CONCLUSIONS

This study across a portion of the Great Valley of West Virginia has shown that soil-gas radon concentrations correlate with lithologic changes of mapped geologic formations and thickness and character of residuum and soil. High concentrations of soil-gas radon occur in relatively deep, clay-rich, red residuum overlying parts of the Elbrook, Conococheague, and Beekmantown Formations and in residuum and soils overlying parts of the Martinsburg Formation. Total count radioactivity on the surface of soil and residuum has similar trends as adjacent soil-gas radon concentrations. Available aeroradioactivity data correlate posi-

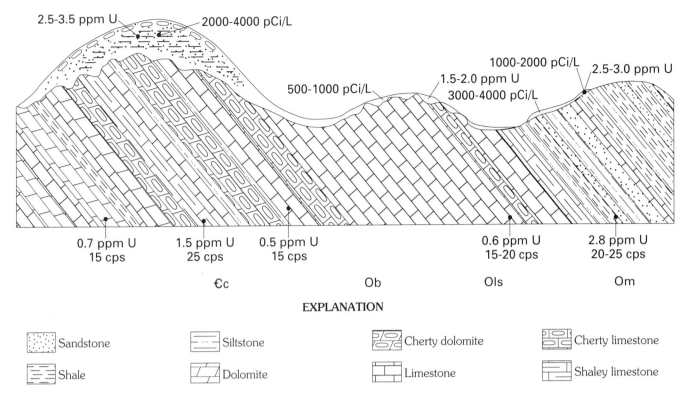

Figure 16. Model for distribution of uranium and soil-gas radon in the Great Valley across the study area (weathering model modified from Hack, 1965). Formation symbols as in Figure 2.

tively with soil gas-radon and surface radioactivity. In some cases, areas of high soil-gas radon may not have been detected by the airborne surveys or ground radioactivity surveys.

No apparent correlation exists between bedrock cleavage, fractures, joints, faults and calcite veins, and uranium concentration or soil-gas radon. Evidence suggests that solution of carbonate bedrock, even with very low levels of uranium, can generate an overlying residuum with 5 to 10 times more uranium content. This residuum, when thick, red, and clay-rich, contains sufficient amounts of uranium and radium to cause high soil-gas radon concentrations.

These results suggest that homes built on carbonate bedrock with thin or nonexistent residuum and soil are more likely to have low indoor radon values. Homes built on thick, red, clay-rich residuum and soil that have formed over lithologies of the Elbrook Conococheague, and Beekmantown Formations are more likely to have indoor radon values exceeding present EPA guidelines. Soil-gas radon concentrations in homes built on bedrock and residuum and soil in areas underlain by the Martinsburg Formation may have values that exceed the EPA action level.

Existing geologic and radiometric data bases are useful in predicting regional trends of radon soil-gas concentrations. In the Great Valley of West Virginia, soil survey maps and thickness of residuum maps are most helpful in outlining areas of high radon soil-gas hazards.

ACKNOWLEDGMENTS

This study was funded in part by the Evolution of Sedimentary Basins Program of the U.S. Geological Survey and by the U.S. Department of Energy. We thank Linda Gundersen of the U.S. Geological Survey for her continuous support throughout this project and as a reviewer on all phases of this chapter. We also thank James Otton of the U.S. Geological Survey for his advice on radon and as a reviewer of early drafts. The final manuscript was greatly improved by the reviews of Mark McKoy of the West Virginia Geological Survey, Stanley Johnson of the Virginia Division of Mineral Resources, and Alexander Gates of the New York Geological Survey. Finally, we thank Nancy Polend of the U.S. Geological Survey for drafting the final figures.

REFERENCES CITED

Brower, S. D., 1991, The distribution and hydrogeologic controls of randon gas in ground water, soil gas, and indoor air in Jefferson and Berkeley Counties, West Virginia: Morgantown, West Virginia University (M.S. thesis), 213 p.

Cardwell, D. H., Erwin, R. B., and Woodward, H. P., 1968, Geologic map of West Virginia: West Virginia Geological and Economic Survey, scale 1:125,000.

Epstein, A. G., Epstein, J. B., and Harris, L. D., 1977, Conodont color alteration— An index to organic metamorphism: U.S. Geological Survey Professional Paper 995, 27 p.

Gabelman, J. W., 1977, Migration of uranium and thorium—Exploration significance: American Association of Petroleum Geologists Studies in Geology, no. 3, Table 3, p. 13.

Gorman, J. L., Pasto, J. K., and Crocker, C. D., 1966, Soil survey of Berkeley County, West Virginia: U.S. Department of Agriculture, Soil Conservation Service, Series 1960, no. 30, 141 p.

Greenman, D. J., Rose, A. W., and Jester, W. A., 1990, Form and behavior of radium, uranium and thorium in central Pennsylvania soils derived from dolomite: American Geophysical Union Research Letters Paper 9L8256, 6 p.

Gundersen, L.C.S., 1989, Predicting the occurrence of indoor radon: A geologic approach to a national problem: EOS Transactions of the American Geophysical Union, v. 70, no. 15, p. 280–281.

Hack, J. T., 1965, Geomorphology of the Shenandoah Valley, Virginia and West Virginia and origin of residual ore deposits: U.S. Geological Survey Professional Paper 484, 84 p.

Hatfield, W. F., and Warner, J. W., 1973, Soil survey of Jefferson County, West Virginia: U.S. Department of Agriculture, Soil Conservation Service, 81 p.

Hearn, P. P., Jr., Sutter, J. F., and Belkin, H. E., 1987, Evidence for Late-Paleozoic brine migration in Cambrian carbonate rocks of the central and southern Appalachians: Implications for Mississippi Valley–type sulfide mineralization: Geochimica et Cosmochimica Acta, v. 51, no. 5, p. 1323–1334.

Neuschel, S. K., 1965, Natural gamma aeroradioactivity of the District of Columbia and parts of Maryland, Virginia, and West Virginia: U.S. Geological Survey Geological Investigations Map GP-475, scale 1:250,000.

Reimer, G. M., 1992, A technique for sampling radon in soil and water, in Gundersen, L.C.S., and Wanty, R. B., eds., Field studies of radon in natural rocks, soils, and water: U.S. Geological Survey Bulletin, in press.

Shumaker, R. C., Wilson, T. H., Dunne, W. M., Knotts, J., and Buckley, R., 1985, Cross sections of Pennsylvania, Virginia and West Virginia, in Woodward, N. B., ed., Valley and Ridge thrust belt: Balanced structure sections, Pennsylvania to Alabama: Appalachian Basin Industrial Association in cooperation with University of Tennessee, Department of Geology Studies in Geology 12, p. 19.

Texas Instruments, Inc., 1978, Aerial radiometric and magnetic reconnaissance survey of Baltimore, Washington, and Richmond Quadrangles: U.S. Department of Energy Contract EY-76-13-1644, GJBX-133(78), 10 p.

MANUSCRIPT ACCEPTED BY THE SOCIETY APRIL 6, 1992

Soil radon distribution in glaciated areas: An example from the New Jersey Highlands

Alexander E. Gates, Lawrence Malizzi, and John Driscoll III
Department of Geology, Rutgers University, Newark, New Jersey 07102

ABSTRACT

In contrast to results of regional soil radon studies in unglaciated areas, bedrock geology shows no correlation with radon concentrations in glacial soils overlying the Green Pond outlier and Reservoir fault zone, New Jersey Highlands. Total gamma radiation and uranium concentrations in the Paleozoic sedimentary rocks of the Green Pond outlier are generally lower than in the Precambrian gneisses of the Reading Prong to the west. The sedimentary bedrock shows average gamma radiation of 220 c/s (136 to 323 c/s) and uranium concentrations of 0.5 to 0.6 ppm, whereas gamma radiation from the Grenville gneisses averages 284 c/s (240 to 576 c/s) and average uranium concentrations are 1.2 to 2.2 ppm. Rare pegmatites that occur along the Reservoir fault zone yield anomalously high average gamma radiation of 2,018 c/s (1,949 to 3,495 c/s) and average uranium concentrations of 28.5 ppm.

Radon concentrations in the glacial soil cover exhibited similar averages with wide ranges regardless of underlying bedrock geology. No appreciable difference was found between soil radon concentrations over Paleozoic sedimentary units, the pegmatites, fault zone, or Precambrian gneisses. Radon from soil over the Paleozoic sedimentary bedrock averaged 518 pCi/L (237 to 2,695 pCi/L), whereas it averaged 527 pCi/L (200 to 1,872 pCi/L) in soil over the Grenville gneisses.

The Green Pond outlier and the Reservoir fault zone are blanketed by the Wisconsin-age glacial cover and contain recessional deposits that are proximal to the terminal moraine. All soil radon was sampled in the glacial cover. The glacial sediments contain erratics primarily composed of lithologies in the area but also of exotic rock types. Because uranium concentrations in the erratics and matrix are highly variable, soil radon of individual samples is governed by the local uranium concentrations. Other possible reasons for the lack of correlation between bedrock and soil radon are the high permeability of the glacial soil that permits radon diffusion, atmospheric dilution, and variations in cover thickness.

INTRODUCTION

Detailed soil radon studies in unglaciated areas show a strong correlation between the bedrock geology and radon in overlying soil (Gates and Gundersen, 1989a, b; Gundersen and others, 1987, 1988a, b; Gates and others, 1990). If all other factors are equal, bedrock with high uranium content produces higher radon concentrations in the overlying soil than bedrock with low uranium content. Rocks in shear zones have higher permeability and radon emanation coefficients than their undeformed counterparts also accounting for the anomalously high soil radon found in these areas (Gates and Gundersen, 1989a; Gundersen and others, 1988a; Gates and others, 1990). In unglaciated areas, the overlying soils commonly reflect the chemistry of the bedrock; therefore, the bedrock controls the radon content.

In this study, soil radon and bedrock geology were investigated in a glaciated area proximal to a terminal moraine to determine if bedrock geology-soil radon correlations can be

---

Gates, A. E., Malizzi, L., and Driscoll, J., III, 1992, Soil radon distribution in glaciated areas: An example from the New Jersey Highlands, *in* Gates, A. E., and Gundersen, L.C.S., eds., Geologic Controls on Radon: Boulder, Colorado, Geological Society of America Special Paper 271.

universally applied. Additional goals were to investigate whether radon is a function of glacial till type, and whether radon can be a potential problem in glaciated areas.

Regional geology

The study area is located in north-central New Jersey and contains the Paleozoic sedimentary rocks of the Green Pond outlier and the Grenville gneisses of the New Jersey Highlands (Fig. 1). The New Jersey Highlands are part of the Reading Prong, which has been shown to contain some of the highest indoor radon concentrations in the country (Smith and others, 1987; Gundersen and others, 1988a). The Paleozoic Green Pond outlier is separated from the Grenville gneisses on its western side by the Reservoir fault zone. Concentrations of uranium in bedrock vary markedly across the area with local anomalies in the Precambrian units. Exxon Inc. conducted uranium exploration in the area and drilled the Reservoir fault zone during the 1970s. Based on analogy with previous radon studies, high uranium concentrations in the bedrock would be expected to produce high radon concentrations in the overlying soil. This area, however, has a major complication: it is proximal to a terminal moraine and is covered by glacial deposits in the form of drumlins, kames, eskers, meltwater channels, and outwash plains.

Stratigraphy. The oldest units in the study area are the Proterozoic gneisses of the New Jersey Highlands (Fig. 2, Table 1). Based on correlation with Canada Hill granite, the gneisses are older than 913 Ma (Helenek and Mose, 1984). The medium-

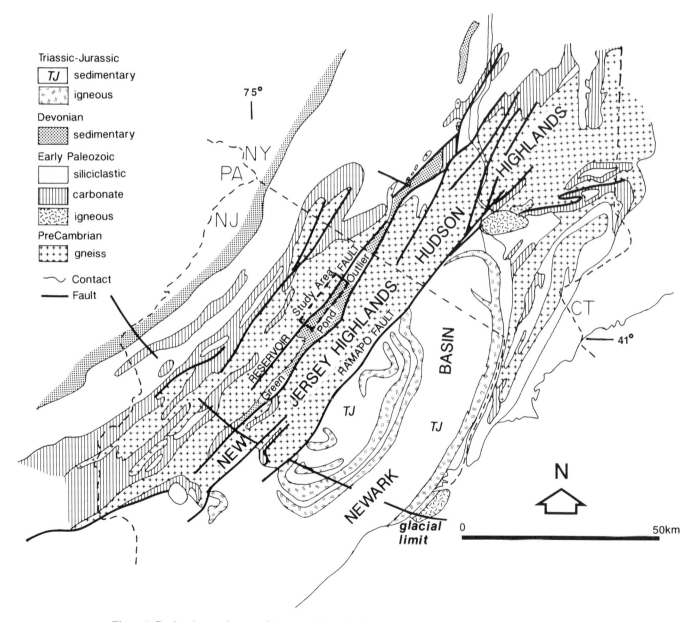

Figure 1. Regional tectonic map of the central Appalachians showing location of the study area (Fig. 2).

grained gneisses consist of biotite, pyroxene, and hornblende quartz monzonite gneiss and graphite, pyroxene, biotite, and hornblende granite to alkali-feldspar granite gneiss.

The gneisses also contain secondary calcite, epidote, chlorite, monazite, and hematite in veins. All exhibit pervasive gneissic foliation and localized cataclastic zones. The quartz monzonite gneiss contains plagioclase, biotite, hornblende or clinopyroxene, quartz, and microcline with accessory magnetite, apatite, garnet, zircon, and titanite. The granite gneisses form four distinct units (Fig. 2): biotite granite gneiss, pyroxene granite gneiss, hornblende granite gneiss, and graphite granite gneiss. All units have similar compositions and differ only in the predominant varietal mafic mineral. Each contains interlayers of alkali-feldspar granite. The granite gneisses contain quartz, plagioclase, and microcline with subsidiary amounts of biotite, augite, hypersthene, hornblende, and graphite with accessory magnetite, ilmenite, zircon, apatite, and sphene. They also contain zones of locally abundant scapolite, tremolite, apatite, potassium feldspar, diopside, plagioclase, and titanite.

Age relations among the different Proterozoic units are unknown, but all the gneisses are cut by rare uraniferous allanite-, biotite-, zircon-, and monazite-bearing granitic pegmatites only along the Reservoir fault. Major minerals of the pegmatites are quartz, plagioclase, and microcline. The minor minerals are biotite and muscovite with accessory zircon and secondary vein-filling monazite associated with hematite and epidote.

The Precambrian gneisses are unconformably overlain by Paleozoic sedimentary rocks (Table 1) of the Green Pond outlier. The Paleozoic rocks comprise an incomplete Ordovician through Devonian sequence that is typical of the Valley and Ridge Province to the west (Kümmel and Weller, 1902). The Ordovician Martinsburg Formation, a black slate, is the oldest unit in the study area. The coarse siliciclastics of the Middle Silurian Green Pond (correlative of the Shawangunk Formation) and the Longwood (correlative of the Bloomsburg Formation) Formations unconformably overlie the Martinsburg Formation (Wolfe, 1977). The Upper Silurian–age units are carbonates represented by the Poxono Island and Berkshire Valley Formations. The overlying Connelly (correlative of the Oriskany Formation), Esopus, and Kanouse Formations are Lower Devonian shales, siltstones, and sandstones. The Middle Devonian units include the black shale and sandstone of the Bellvale Formation and red conglomerate of the Skunnemunk Formation (correlative of the Catskill Formation).

Glacial cover. The Reservoir fault and Green Pond outlier are covered with Wisconsin-age glacial deposits (Walters, 1975).

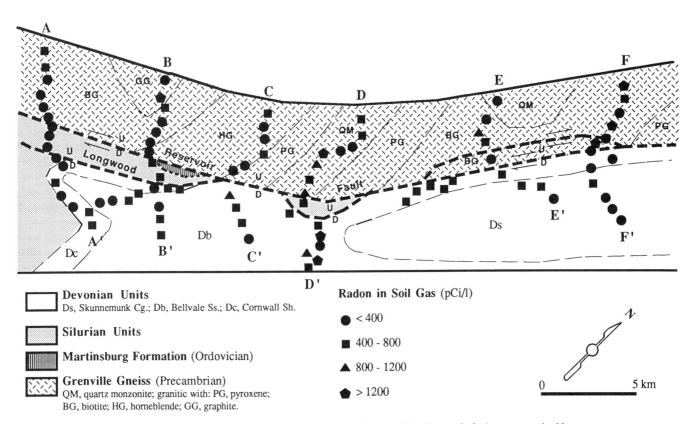

Figure 2. Geologic and soil gas radon map of the study area. The Reservoir fault separates the New Jersey Highlands/Reading Prong to the northwest from the Green Pond outlier to the southeast (modified after Herman and Mitchell, 1991). Traverses as shown; see text for explanation. Light dashed lines indicate contacts; heavy dashed lines, faults with movement senses as indicated.

TABLE 1. STRATIGRAPHY OF THE STUDY AREA
DEVONIAN
Skunnemunk Conglomerate Thin- to thick-bedded, quartz pebble conglomerate with a red medium-grained sandstone matrix. Interbeds of medium-grained red sandstone. Conglomerate is locally crossbedded. 915 m thick
Bellvale Sandstone Thin- to thick-bedded gray, medium-grained sandstone interbedded with black shale. Locally fossiliferous and crossbedded. 600 m thick
Cornwall Shale Thin-bedded, fine-grained fissile black shale interlayered with laminated gray siltstone. Moderately fossiliferous. 300 m thick
Kanouse Sandstone Medium- to thick-bedded, gray to tan conglomerate and coarse to fine-grained graded sandstone 15 m thick
Esopus Formation Thin interlayered gray mudstone and medium-grained sandstone. Fossiliferous. 60–100 m thick
Connelly Conglomerate Quartz pebble conglomerate with a tan, medium-grained sandstone matrix. 12 m thick
SILURIAN
Birkshire Valley Formation Thin-bedded limestone with interlayers of gray, intraformational dolomitic breccia. Thickness unknown
Poxono Island Formation Medium-bedded gray dolomite interlayered with thin bedded medium-grained calcareous sandstone. 80–130 m thick
Longwood Shale Purple shale with interlayers of red cross-bedded medium-grained sandstone. 100 m thick
Green Pond Conglomerate Quartz pebble conglomerate with a medium-grained sandy matrix and silica cement. The conglomerate contains medium-bedded interlayers of cross-bedded sandstone. 300 m thick
ORDOVICIAN
Martinsburg Shale Black, slaty shale with thin beds of medium-grained sandstone. Moderately fossiliferous and cross-bedded. Thickness unknown
PRECAMBRIAN
Granite Pegmatites Very coarse grained quartz, plagioclase, and microcline pegmatite dikes with minor biotite. Accessory zircon and apatite. Secondary epidote, hematite, and monazite. 1–10 m thick
Quartz Monzonite Gneisses Medium-grained hornblende, pyroxene, and biotite, quartz monzonite gneisses with quartz, plagioclase (An_{37}), and microcline. Accessory magnetite, apatite, garnet, sphene, and zircon with secondary chlorite
Granite Gneisses Medium-grained hornblende, pyroxene, graphite, and biotite granite to alkali-feldspar granite gneisses with microcline and plagioclase (An_{38}). Accessory sphene, apatite, zircon, and magnetite with secondary epidote, chlorite, monazite, and hematite. Locally contains abundant scapolite, tremolite, and titanite

The glacial cover in the study area consists of upland till and valley deposits (Shenker and Caldwell, 1976). The upland till consists primarily of unstratified soils as thick as 10 m on hilltops but it is thin to nonexistent on slopes. The till contains lesser amounts of stratified drift and meter-scale erratics in a poorly sorted and very porous and permeable sandy matrix (Wolfe, 1977; S. Stanford, personal communication, 1989). The composition of the erratics is highly variable and consists primarily of local rock types with few exotics. Most erratics in the till are composed of quartz monzonite and granite gneiss, granitic pegmatite, and sandstone and conglomerate from the sedimentary units of the Green Pond outlier. The valley glaciofluvial deposition occurred during glacial recession in the form of kames, outwash plains, meltwater channels, and eskers (Lewis and Kümmel, 1912; Stanford, 1990). These glacial features consist of stratified sand and gravel as thick as 30 m, with an assortment of cobbles consisting primarily of local rock types.

METHODS OF STUDY

Soil radon concentrations were determined from grab samples of soil gas, and gamma radiation was measured directly on the rock and soil surfaces. The radon and gamma radiation data were collected along six 3-km-long transects that cross the Reservoir fault and extend into both the Grenville gneiss and the sedimentary rocks of the Green Pond outlier (Fig. 2). Transects were spaced at 5-km intervals along the 30-km-long mapped trace of the fault zone, and samples were collected at 0.2-km spacings. Along each transect gamma radiation was measured on outcrops of the Grenville gneisses and on Paleozoic sedimentary rocks, as well as on the glacial cover where the soil gas was sampled. The soil gas was sampled by inserting a 75-cm hollow steel probe into the soil and withdrawing 20 ml of gas (method described in Gates and Gundersen, 1989a, and Reimer, 1990). Radon concentrations in the soil gas samples were determined in an evacuated scintillation chamber of an alpha scintillometer using three 2-min analyses. The alpha scintillometer expresses radon in counts that are then converted to picoCuries/liter.

Gamma radiation reflects total radiogenic uranium, thorium, and potassium of the rock as well as background radiation and is expressed in counts/second. Whole-rock uranium, thorium and radium concentrations were determined for 20 representative samples from the units within the field area. Both fluorimetric and spectrographic methods were used in the analysis, which was performed by Geolabs, Golden, CO.

DISTRIBUTION OF RADIOACTIVE ELEMENTS

The Precambrian gneisses exhibit generally higher gamma radiation and uranium concentrations than do the Paleozoic sedimentary units, and the pegmatites exhibit the highest gamma radiation and uranium concentrations (Fig. 3, left and center). Sedimentary bedrock exhibits relatively low gamma radiation and uranium concentrations. The Devonian units exhibit slightly

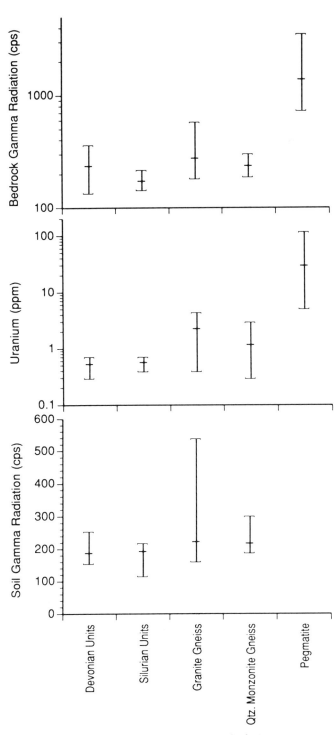

Figure 3. Plots of lithologies in the Reservoir fault area vs.: top, log-scaled gamma radiation of bedrock; center, log-scaled uranium concentrations of bedrock; bottom, gamma radiation of soil above bedrock. Vertical bars show ranges; horizontal cross bars show averages.

higher average gamma radiation and locally higher uranium concentrations than do the Silurian units. The elevated radioactivity in the Devonian units is because these rocks include shale that is more radioactive than the sandstone and conglomerate in the Silurian units. Gamma radiation and uranium concentrations in the gneisses are generally higher than the sedimentary units but are variable. Mafic and calc-silicate gneisses are generally less radioactive than average. Alkali-feldspar granite gneisses are the most radioactive of these rocks, and contain the highest concentrations of uranium. The rare radioactive pegmatites dikes exhibit anomalously high average gamma radiation and uranium concentrations. These dikes are easily identified using a gamma scintillometer.

The range of gamma radiation of all soils exhibits a nearly complete overlap throughout the area (Fig. 3, right). Gamma radiation in the glacial cover over the gneisses averages 220 c/s, whereas the glacial cover over the sedimentary units averages 199 c/s. Radioactivity varies greatly from location to location, depending on the till composition and thickness at the sample station. Because the till is a mixture of the bedrock types, any level of gamma radiation within the range of the field area is possible regardless of the bedrock beneath the glacial deposits. In general, the radioactivity of the soils is uniform over the area. Where soil cover is thin, underlying bedrock contributes to the radioactivity. In these areas there is a slight correlation between bedrock geology and gamma radiation.

SOIL RADON

Soil radon was evaluated from within the glacial cover over both the Grenville gneisses and Paleozoic sedimentary rocks (Figs. 2, 4a). Six traverses of radon in soil crossed from the Reading Prong gneisses into the Paleozoic sedimentary rocks. A total of 110 soil gas samples were taken, including 15 duplicates. Using these data, not even major bedrock boundaries could be identified. In most traverses (see especially A–A′), there was little to no change in radon concentrations across the Reservoir fault. In traverses B–B′ and C–C′, the soil gas radon concentrations were higher over the sedimentary rocks than the gneisses. In traverse F–F′, samples with the highest and lowest radon concentrations of the traverse occurred adjacent to each other in several places. The highest radon concentration, 2,695 pCi/L, was found in soil over Devonian sandstone that exhibits low gamma radiation and uranium concentrations (Fig. 2). The lowest radon concentration, 200 pCi/L, occurred over Precambrian biotite granite gneiss that exhibits relatively high gamma radiation and uranium concentrations. Low radon concentrations were present in the soil overlying granitic pegmatites that exhibit anomalously high gamma radiation. In general, radon concentrations in soils over gneisses is virtually the same as that in soils over the sedimentary bedrock of the Green Pond outlier. Radon concentrations in the soil yielded similar averages with wide ranges regardless of the underlying bedrock.

Soil radon concentrations show minor correspondence with

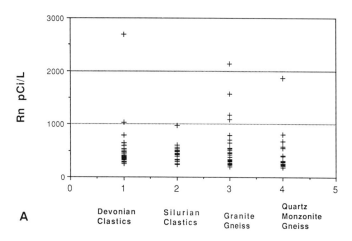

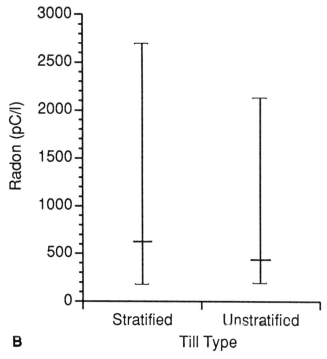

Figure 4. Graphs of radon in soil gas vs.: A. Bedrock lithology. B. Glacial till type. See text for explanation of till type.

glacial sediment type. Radon concentrations are higher in glaciofluvial deposits than in the unstratified till (Fig. 4b). Radon in the glaciofluvial deposits averaged 638 pCi/L, whereas radon in the unstratified till averaged 438 pCi/L. Although the average radon concentrations appear to indicate lithologic control, the concentrations in individual samples show a broad overlap through most of the range. Radon concentrations vary greatly from sample to sample and appear to reflect local conditions rather than a general

control. The Reservoir fault–Green Pond outlier area, however, contains few types and relatively similar glacial sediments because of its proximity to the terminal moraine. Additional work is required to address the control of glacial lithologies on radon in detail.

DISCUSSION

Soil radon concentrations do not reflect the bedrock geology in the Reservoir fault and Green Pond outlier but are probably locally controlled by the composition of the clasts in the glacial cover and cover thickness. Soils over the gneisses exhibit a wide range both in radon concentrations and in gamma radiation (Fig. 5). Conversely, soil over the sedimentary units exhibits a similarly wide range in radon concentrations but a narrow range in gamma radiation. Average collective soil radon concentrations over the Grenville gneisses (527 pCi/L) and Paleozoic sedimentary rocks (518 pCi/L) are virtually the same.

In unglaciated areas, where the previous regional radon studies have been focused (Gundersen and others, 1988a, b; Gates and Gundersen, 1989a, b), the soil is derived directly from the underlying bedrock. These soils are composed of altered bedrock and therefore their compositions reflect those of the underlying rock. The radon produced by the soil is a function of the uranium content of the bedrock. These soils additionally tend to be clay-rich and of relatively low permeability. In such soils, radon transport distances are short before decay occurs. Ambient soil radon is therefore not significantly affected by atmospheric dilution or concentration gradients in these soils.

The composition of the local glacial cover is the major factor that controls soil gas radon distribution in glaciated areas. The composition of both the till and glaciofluvial deposits is relatively homogeneous on the map scale but highly variable on the clast scale. Randomly distributed erratics with high uranium concentrations locally produce high radon concentrations in the surrounding soil. In most of the area, however, the uranium content

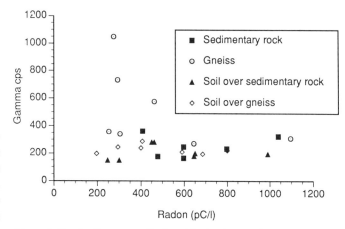

Figure 5. Graph of gamma radiation of bedrock and soil vs. radon in soil gas where measured simultaneously.

of the till in any given location is relatively low and produces lesser amounts of radon. Another factor that controls the radon distribution in the area is the porosity and permeability of the soils. The upland-type glacial tills are clay-poor and highly permeable. Poor packing of boulder- and cobble-rich tills creates well-interconnected pore spaces as large as 20 cm in diameter. Increased permeability leads to more radon diffusion and atmospheric dilution (Tanner, 1980). The result is that much of the radon is able to escape to the atmosphere or mix with that produced in adjacent soil. The soil radon concentrations are therefore lower than would be expected, considering the uranium concentrations of the rocks, and more uniform than would be expected, considering the distribution of the uranium-bearing rocks. The stratified glaciofluvial deposits are less permeable than the unstratified till. They are finer grained, contain more clay, and exhibit a more uniform porosity. The stratification provides an impedance to vertical radon migration. The slight correlation between relatively higher soil radon and glaciofluvial deposits appears to be the result of reduced permeability rather than composition.

CONCLUSIONS

In contrast to the results of radon studies in unglaciated terranes, this study found no correlation between bedrock geology and radon in overlying soil in the Reservoir fault and Green Pond outlier, a glaciated area proximal to a terminal moraine. The Reservoir fault separates relatively uranium-rich gneisses of the Reading Prong, including uraniferous pegmatites, from uranium-poor sedimentary rocks of the Green Pond outlier. Based on the results of previous regional radon studies, this area would be expected to produce a distinctive pattern of radon concentrations in the soil. The area, however, is covered by Wisconsin-age glacial sediments in contrast to previous study areas where residual soils are derived directly from bedrock. The radon was nearly homogeneously distributed across the Reservoir fault and Green Pond outlier area with local randomly distributed anomalies. We conclude that soil radon concentrations in glaciated areas are controlled by local glacial cover thickness, permeability, and porosity, as well as by the composition of the local till clasts.

This study indicates that local radon surveys must be conducted in order to identify areas of potentially high soil radon. Generalizations about radon distribution based on studies in unglaciated areas cannot be applied to glaciated areas proximal to terminal moraines. There may, however, be minor control on soil radon by the type of glacial deposits.

ACKNOWLEDGMENTS

Laboratory and field work for this project was funded through the U.S. Geological Survey from Grant 05-88ER60665 from the U.S. Department of Energy (to Linda Gundersen and A. G.). Thanks are given to Greg Herman, James Mitchell, and Scott Stanford for useful discussions. The reviews of R. Fakinduny, L.C.S. Gundersen, W. Manspeizer, K. Muessig, P. Muller, J. H. Puffer, and A. H. Vassiliou substantially improved the text.

REFERENCES CITED

Gates, A. E., and Gundersen, L.C.S., 1989a, Effect of ductile shearing on soil radon in the Brookneal Zone, VA: Geology, v. 17, p. 391–394.
—— , 1989b, Radon distribution around the Hylas Zone, VA.: A product of lithology and ductile shearing: Geological Society of America Abstracts with Programs, v. 21, p. 17.
Gates, A. E., Gundersen, L.C.S., and Malizzi, L. D., 1990, Comparison of radon in soil over faulted crystalline terranes: Glaciated versus unglaciated: Geophysical Research Letters, v. 17, p. 813–816.
Gundersen, L.C.S., Reimer, G. M., and Agard, S. S., 1987, Geologic control of radon in Boyertown and Easton, PA: Geological Society of America Abstracts with Programs, v. 19, p. 87.
—— , 1988a, The correlation between geology, radon in soil gas, and indoor radon in the Reading Prong, in Marikos, M., ed., Proceedings of the GEO-RAD conference: Geological causes of radionuclide anomalies: Missouri Department of Natural Resources Special Publication 4, p. 99–111.
Gundersen, L.C.S., Reimer, G. M., Wiggs, C. R., and Rice, C. A., 1988b, Radon potential of rocks and soils in Montgomery County, Maryland: U.S. Geological Survey Miscellaneous Field Study 2043.
Helenek, H. L., and Mose, D. G., 1984, Geology and geochronology of Canada Hill Granite and its bearing on the timing of Grenvillian events in the Hudson Highlands, New York, in Bartholemew, M. J., ed., Grenville events and related topics in the Appalachians: Geological Society of America Special Paper, v. 194, p. 57.
Herman, G. C., and Mitchell, J. P., 1991, Geology of the Green Pond outlier from Dover to Greenwood Lake, New Jersey: New Jersey Geological Survey Open File Map, Department of Environmental Protection.
Kümmel, H. B., and Weller, S., 1902, The rocks of the Green Pond mountain region: New Jersey Geological Survey Annual Report of the State Geologist 1901, p. 1–51.
Lewis, J. V., and Kümmel, H. B., 1912, Geologic map of New Jersey (1910–1912): New Jersey Department of Conservation and Economic Development, Atlas Sheet 20.
Reimer, G. M., 1990, Reconnaissance techniques for determining soil-gas radon concentrations: An example from Prince Georges County, Maryland: Geophysical Research Letters, v. 17, p. 809–812.
Shenker, A., and Caldwell, D. H., 1976, Environments associated with the Wisconsin terminal moraine between Belvidere and Netcong, N.J.: New Jersey Academy of Sciences Bulletin, v. 21, p. 26–27.
Smith, R. C., II, Reilly, M. A., Rose, A. W., Barnes, J. H., and Berkheiser, S. W., 1987, Radon: A profound case: Pennsylvania Geology, v. 18, p. 1–7.
Stanford, S., 1992, Surficial geology of the Newfoundland Quadrangle, N.J.: N.J. Geological Survey Open File Map, Department of Environmental Protection: (in press).
Tanner, A. B., 1980, Radon migration in the ground: A supplementary review, in Gesell, T. F., and Lowder, W. M., eds., Natural radiation environment III, Symposium Proceedings 1, p. 5–56.
Walters, J., 1975, Origin and significance of fossil periglacial phenomena in central and southern New Jersey: New Brunswick, New Jersey, Rutgers University (Ph.D. thesis), 176 p.
Wolfe, P. E., 1977, The geology and landscapes of New Jersey: Crane and Russak, p. 70–200.

MANUSCRIPT ACCEPTED BY THE SOCIETY APRIL 6, 1992

Radon in the Coastal Plain of Texas, Alabama, and New Jersey

Linda C. S. Gundersen
U.S. Geological Survey, Box 25046, Federal Center, MS-939, Denver, Colorado 80225
R. Thomas Peake
U.S. Environmental Protection Agency, Washington, D.C. 24060

ABSTRACT

In comparison with other geologic terranes, the Coastal Plain of the eastern and southern United States is a region of low to moderate radon potential with local areas of high radon potential. Analyses were made of soil radon concentrations, soil permeability, and equivalent uranium concentrations. Descriptions of soil profiles include grain size analyses, relative moisture concentrations, and mineralogy. Data from Texas, Alabama, and New Jersey indicate that Cretaceous and lower Tertiary glauconitic sands, Cretaceous and Tertiary chalks, carbonaceous shales, and phosphatic sediments have the highest radon-producing potential. Marine limestones and quartz sands have the lowest radon-producing potential. More than half the soil radon concentrations measured were less than 500 pCi/L; however, 20% of the soil radon concentrations were 2,000 pCi/L or more. The highest soil radon concentration sampled was 16,200 pCi/L measured in the glauconitic sands of the Navesink Formation in New Jersey.

INTRODUCTION

The Coastal Plain of the eastern and southern United States comprises a systematic progression of predominantly marine and fluvial sediments deposited during the evolution of the Atlantic and Gulf Coasts. The oldest rocks and sediments exposed in the Coastal Plain are Cretaceous in age and generally consist of glauconitic sandstones, sands, chalks, and clays, as well as nonglauconitic quartz sandstones, sand, clays, and fossiliferous limestone. These are overlain by lower Tertiary (Paleocene, Eocene, and Oligocene) sands and clays, often glauconitic, carbonaceous, or locally phosphatic, and upper Tertiary (Miocene) fossiliferous chalks, clays, and thin sands. The youngest Tertiary sediments (Pliocene) are dominated by gravelly sands, clayey sands, and thin clay beds. Because of the consistency in the general stratigraphy of the Coastal Plain, many of the lithologic sequences are regionally similar from state to state.

Data from indoor radon surveys (Cohen and Gromicko, 1988; Ronca-Battista and others, 1988; Cohen, 1990; Dziuban and others, 1990) indicate that the Atlantic and Gulf Coastal Plains are generally regions of low indoor radon but can have moderate and locally high levels (>4 pCi/L). An aerial radiometric map of the United States compiled from the National Uranium Resource Evaluation Program by Duval and others (1989) indicates that the Cretaceous and upper Tertiary sediments commonly have high radioactivity [>2.5 equivalent uranium (eU) ppm] in relation to surrounding sediments. In this study, several geologic and geochemical surveys were made to identify specific characteristics associated with the different levels of radon potential and to develop a general predictive model for radon in the Coastal Plain. Survey transects perpendicular to geologic strike were chosen in the Coastal Plain of New Jersey, Alabama, and Texas. Along each of these transects, radon in soil gas, surface gamma-ray activity, and permeability were measured, and core and auger samples of soils and sediment were examined.

METHODS

Three hundred soil gas samples were taken at intervals of 2.5 to 3.0 km using the Reimer grab sampling technique (RGS). A more detailed description of this technique is given by Reimer (1991). A steel probe with an 8-mm diameter is driven into the ground to a depth of 1 m. A grab sample of 20 cm³ is withdrawn through an air-tight septum in a removable cap fitted on top of the probe. Soil air enters the probe through 10 holes that pierce a recessed zone of the probe 1.3 cm from the end. The sample is measured 4 to 8 hr later in an alpha scintillometer. In this study, a Pylon alpha scintillometer and Pylon scintillation cells were used. At every fourth sampling site, the sample hole was reoccupied, and in situ, flow-through radon and permeability were measured using the MKII Radon/Permeability Sampler developed by Rogers and Associates. A total of 64 sites were sampled using both the RGS and MKII techniques. The MKII employs a probe of comparable design to the RGS, except the diameter is 1.3 cm and 20 air entry holes are located near the bottom of the probe in a recessed zone. The probe is attached to a series of vacuum gauges, an air pump, a flow meter, and a Pylon alpha scintillometer. Air flow is established in the probe through suction on the soil. Permeability is measured first and then the gas sample is passed through a scintillation cell in the alpha scintillometer. A more detailed description of the technique and equipment is given in Nielson and others (1989). At each sample site, soil grain size and moisture conditions were also recorded. At 10% of the sampling sites, duplicate samples were collected and analyzed to ensure quality control. No discrepancies were found in the duplicate samples. Surface gamma measurements were made at all the sample sites in New Jersey and Alabama with a GAD-6 portable gamma-ray spectrometer that provides equivalent uranium in parts per million. Surface gamma measurements were made at only 23 of the sites in Texas. These measurements are designated "equivalent" uranium because they are actually calculated from the spectral abundances of ^{214}Bi, a daughter product of ^{238}U. The calculation assumes secular equilibrium between uranium and its daughter products.

TRANSECT RESULTS

Figure 1 shows the average soil radon concentrations along each transect, from east to west in New Jersey, from north to south in Alabama, and from west to east in Texas. Average values of soil radon concentrations were calculated for each geologic unit with two or more successfully measured samples using the RGS technique. The plots are aligned with each other by geology to show the similarity of the radon concentrations from transect to

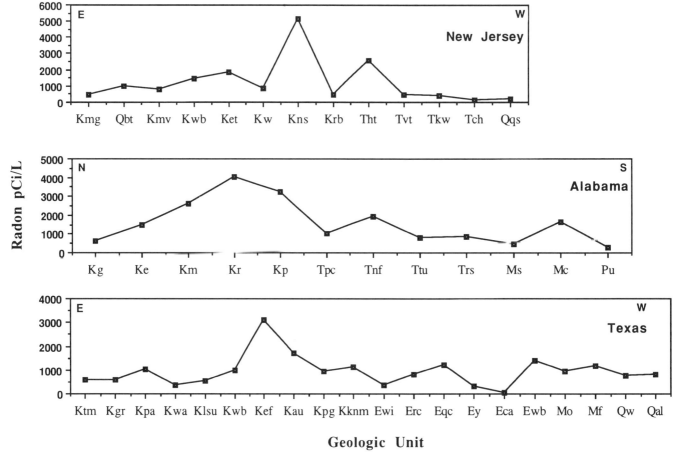

Figure 1. Plots of average soil-gas radon by geologic unit in the Coastal Plain of New Jersey, Alabama, and Texas.

transect. There is an obvious strong control on the radon concentration by rock type that is consistent throughout the Coastal Plain. A list of geologic units and descriptions for each state is given in Tables 1 through 3. There is a positive correlation between moderate (1,001 to 2,000 pCi/L) and high (>2,000 pCi/L) soil radon concentrations and the occurrence of glauconitic sandstones, glauconitic clays, marly or phosphatic clays, carbonaceous sands, clays, shales, and chalks. Low concentrations of soil radon average less than 1,000 pCi/L and were commonly measured in quartzose sands, clays, limestones, and shales. These designations of high, moderate, and low radon concentration are used throughout the text and indicate the ranges of soil radon concentrations just given. Equivalent uranium data correlate positively with radon in each of the sediment types. Grain size and permeability exhibit a complex relation with radon and are discussed at the end of the chapter in context of the soil gas measured using the MKII sampler. The following is a summary of the soil radon measured using the RGS technique and compared with the equivalent uranium concentration and the general composition of the sediment.

New Jersey

Two transects were conducted in New Jersey, one in the northern part of the Coastal Plain from Plainsboro to Neptune City and the other toward the southern part of the Coastal Plain from Four Mile Circle to Barnegat (Fig. 2). Data were acquired over a 4-day period in July 1988 during stable, dry hot weather. Soils were generally dry to slightly moist.

The highest soil radon concentrations and eU were found in the glauconitic sands of the Cretaceous Englishtown and Navesink Formations, the Mount Laurel Sand, and the Tertiary Hornerstown Sand. Geologic units that have the lowest soil radon concentrations and equivalent uranium include the clayey, silty quartz sands of the Cretaceous Red Bank Sand, the quartz sands of the Cretaceous Magothy Formation, the Tertiary Kirkwood Formation, the Cohansey Sand, and the Pleistocene residuum. Moderate soil radon concentrations and eU were measured in quartz sands of the Cretaceous Wenonah Formation and the Quaternary(?) Bridgeton Formation, the silty clays of the Cretaceous Woodbury Clay, and the glauconitic sand of the Tertiary Vincentown Formation.

A line graph of soil radon plotted against eU concentration shows a positive correlation (Fig. 3). The graph also displays the distribution of radon and equivalent uranium concentrations by composition and sediment type. Overall, the glauconitic sands have higher soil radon than the quartzitic sands. The six clay samples depicted on the graph are different compositionally and the only clays from which samples were obtained in New Jersey. They include clays from the Merchantville Formation, the Woodbury Clay, and the Kirkwood Formation. The clays all yielded high soil radon concentrations possibly due to the locally phosphatic and glauconitic composition of the different clays; however, the specific composition was not discernable in the field. Despite their low permeability, clays also have high radon emanation because of their high specific surface area. Where clays are dry and fractured, permeability can be very high (Schumann and others, 1990), therefore increasing their overall soil radon emanation. These clays were generally dry with fine fractures in the first meter of soil.

Alabama

The Alabama transect, conducted during April 1989, extends from just north of Montgomery, Alabama, to just south of De Funiak Springs, Florida (Fig. 4). Meterologic conditions were relatively stable, warm, and humid, and soils were generally dry to slightly moist. Short thundershowers occurred on the third and fourth day of the 5-day sampling period. A resampling of the last two stations was conducted following the rain showers to examine any changes in radon concentration in the soil gas. No increase nor decrease in soil radon greater than 15% was observed. It is assumed that rainfall did not significantly affect the soil gas in this instance, as an increase in moisture was not found at the sampling depth.

The highest soil radon concentrations and eU in Alabama were measured in the Cretaceous Mooreville Chalk, carbonaceous sands and clays of the Providence Sand, and the glauconitic sands of the Eutaw and Ripley Formations. Geologic units that have the lowest soil radon concentrations and eU include the quartz sands of the Cretaceous Gordo Formation and quartz sands and residuum of the undifferentiated upper Tertiary sediments. Low to moderate soil radon concentrations and eU were measured in the glauconitic sands and clays of the Tertiary Porters Creek Formation and in the glauconitic sands, limestones, and clays of the Tertiary Nanafalia, Shoal River, and Lisbon Formations, as well as the Tuscahoma Sand.

A line graph of soil radon plotted against eU concentration for the Coastal Plain of Alabama shows a high positive correlation (Fig. 5). Quartzitic sands generally have the lowest soil radon concentrations, and the chalks and carbonaceous sediments have moderate to high soil radon concentrations. The majority of the glauconitic sediments have soil radon measurements >1,000 pCi/L; however, several of the glauconitic sediment samples have low soil radon.

Texas

Two transects, the first through the Cretaceous sedimentary rocks south of the Dallas–Fort Worth area and the second between south of Austin toward Houston, were conducted during April 1988 (Fig. 6). The samples were collected over a period of dry weather and soils were generally dry to slightly moist. A monitor station was maintained where radon was collected and

TABLE 1. GEOLOGIC UNITS AND SOIL RADON IN THE NEW JERSEY COASTAL PLAIN

Geologic Unit	Lithology	# S	# N	Min.	Max
Magothy Fm. (Km)	Sand, quartz, fine to medium grained; local clay beds	4	0	368	568
Merchantville Fm. (Kmv)	Clay, glauconitic, micaceous; local fine glauconitic sand	1	0	771	771
Woodbury Clay (Kwb)	Clay, micaceous, silty	4	0	1,282	1,708
Englishtown Fm. (Ket)	Sand quartz, fine to medium grained; local phosphatic clay	5	0	571	4,458
Marshalltown Fm.	Clay, silty, glauconitic; and sand, quartz	0	0	—	—
Wenonah Fm. (Kw)	Sand, fine grained, silty, slightly glauconitc	3	1	353	1,342
Mount Laurel Sand (Kml)	Sand, quartz, fine to coarse grained, slightly glauconitic	2	1	3,604	3,604
Navesink Fm. (Kns)	Sand, glauconitic, clayey, silty, medium to coarse grained	4	0	419	16,226
Red Bank Sand (Krb)	Sand, quartz, slightly glauconitic, fine to coarse grained, clayey, micaceous	4	0	214	720
Vincentown Fm. (Tvt)	Sand, quartz, glauconitic, fine to coarse grained, clayey, fossiliferous	7	1	106	1,162
Manasquan Fm.	Not sampled; clay, silty, sandy, glauconitic; sand, fine grained	0	0	—	—
Piney Point and Shark River Fms.	Not exposed; clay; sand, quartz, glauconitic	0	0	—	—
Kirkwood Fm. (Tkw)	Sand, quartz, micaceous, fine to medium grained; local phosphatic clay	13	0	54	686
Cohansey Sand (Tch)	Sand, quartz, medium to coarse grained, pebbly; local clay beds	8	0	66	280
Pleistocene residuum (Qqs)	Sand, quartz, clayey, pebbly	2	0	59	280
Bridgeton Fm. (?Qbt)	Sand, quartz, clayey, pebbly	4	0	690	1,293

Note: Letters in parentheses are the unit abbreviations used in Figure 1, with K indicating Cretaceous, T indicating Tertiary, and Q indicating Quaternary. #S is the total number of sample stations in each geologic unit, and # N is the number of stations where samples were not obtained due to impermeable soil. Min. and Max. refer to the minimum and maximum soil radon concentration in picoCuries per liter. Geologic descriptions, unit names, and unit abbreviations were derived from Zapecza (1989) and Owens and Minard (1975). Geologic descriptions were also derived from our own field work. Units are listed from oldest to youngest.

TABLE 2. GEOLOGIC UNITS AND SOIL RADON IN THE ALABAMA COASTAL PLAIN

Geologic Unit	Lithology	# S	# N	Min.	Max.
Fm. (Kck)	Sand, quartz, fine to medium to very coarse grained, pebbly; local clay beds	1	0	1,342	1,342
Gordo Fm. (Kg)	Sand, quartz, fine to medium to very coarse grained, pebbly; local clay beds	5	0	425	801
Eutaw Fm. (Ke)	Sand, quartz, partly glauconitic, fine to medium grained; local carbonaceous clay	7	1	542	2,441
Mooreville Chalk (Km)	Chalk, sandy, fossiliferous	7	5	1,350	3,878
Demopolis Chalk (Kd)	Chalk, micaceous, marly, sandy, fossiliferous	6	6	—	—
Ripley Fm. (Kr)	Clay, micaceous, fossiliferous, glauconitic, sandy; with local thin sands	1	0	4,085	4,085
Providence Sand (Kp)	Sand, quartz, fine to coarse grained; with carbonaceous sand and clay, micaceous, fossiliferous	4	2	2,410	4,011
Clayton Fm. (Tcl)	Limestone, sandy, fossiliferous, with clay, calcareous, silty; and sand, quartz, medium to coarse grained	2	1	286	286
Porters Creek Fm. (Tpc)	Sand, micaceous, clayey, fine to medium grained, glauconitic in part; with clay and limestone, fossiliferous	8	1	606	2,170
Nanafalia Fm. (Tnf)	Sand, quartz, glauconitic in part, fine to coarse grained; and clay, micaceous	5	2	1,205	2,714
Tuscahoma Sand (Ttu)	Sand, quartz, glauconitic, with silt and clay; fine to coarse grained interbeds, carbonaceous, fossiliferous in part	2	2	—	—
Lisbon Fm. (Tl)	Sand, glauconitic clayey, fossiliferous, fine to coarse grained; with sandy limestone	1	0	638	638
Tertiary residuum (Trs)	Clay, sandy; and sand, medium to coarse grained, limonitic, chert and limestone pebbles	21	4	131	1,793
Shoal River Fm. (Ms)	Sand, quartz, silty to fine grained; and clay, sandy	15	1	194	2,308
Citronelle Fm. (Mc)	Gravel, sand, quartz, limonitic; and clay, micaceous	3	2	1,029	1,029
Undifferentiated Pliocene sand (Pu)	Sand, quartz, micaceous, fine to medium grained	2	0	230	363

Note: Letters in parentheses are the same unit abbreviations used in Figure 1, with K indicating Cretaceous, T indicating Tertiary, M indicating Miocene, and P indicating Pliocene. Geologic descriptions, unit names, and unit abreviations were derived from Copeland (1988). Geologic descriptions were also derived from our own field work. Units are listed from oldest to youngest.

TABLE 3. GEOLOGIC UNITS AND SOIL RADON IN THE TEXAS COASTAL PLAIN

Geologic Unit	Lithology	# S	# N	Min.	Max
Twin Mountains Fm. (Ktm)	Sandstone, quartz, fine to coarse grained; interbedded with silty claystone	9	0	181	1,095
Glen Rose Fm. (Kgr)	Interbedded sandy limestone and claystone	11	1	234	1,219
Paluxy Fm. (Kpa)	Sandstone, quartz, fine grained; with interbedded sandy claystone	5	0	340	1,128
Walnut Fm. (Kwa)	Interbedded sandy limestone, fossiliferous; and claystone, calcareous	2	0	161	636
Lower Cretaceous limestone undifferentiated (Klsu)	Interbedded sandy limestone, fossiliferous, locally dolomitic; and claystone, calcareous, marly	9	0	93	1,120
Woodbine Fm. (Kwb)	Sandstone, quartz, limonitic in parts; some clay and shale, fossiliferous	11	0	358	2,126
Eagle Ford Group (Kef)	Shale, carbonaceous; sandstone, quartz, medium grained; and sandy limestone	8	4	1,324	6,333
Austin Chalk (Kau)	Chalk; marl; clay, calcareous, bentonitic	9	1	472	3,725
Ozan Fm. (Ko)	Clay, calcareous, glauconitic, phosphatic, silty, sandy	9	8	1,145	1,145
Pecan Gap Chalk (Kpg)	Marl and clay, sandy, silty	3	1	645	1,228
Marlbrook Marl and Navarro Group (Kknm)	Clay, calcareous to glauconitic, silty	4	0	935	1,548
Wilcox Group (Ewi)	Mudstone, carbonaceous, glauconitic; with sandstone, medium to fine grained	5	2	176	472
Reklaw Fm. and Carrizo Sand (Erc)	Sandstone, quartz, limonitic, medium to coarse grained; and clayey sandstone, glauconitic, micaceous	2	0	272	1,316
Queen City Sand (Eqc)	Sandstone, quartz, limonitic, medium to fine grained; thin clay interbeds	3	0	1,070	1,458
Weches Fm. (Ew)	Sand, glauconitic, marly, fossiliferous; interbedded clay, silty, glauconitic	1	0	817	817
Sparta Sand (Es)	Sand, micaceous, carbonaceous, limonitic, fine to very fine grained; clay partings	11	0	899	899

measured periodically to check the consistency of the radon concentration in the soil over time. Deviations of soil radon concentration greater than 10% were not observed.

The highest soil radon concentration and eU were found in the carbonaceous shales and mudstones of the Cretaceous Eagle Ford, the sands and clays of the Woodbine Formation, and the clay, chalk, and marl of the Austin Chalk. Low to moderate soil radon concentrations and parts per million of equivalent uranium were measured in the quartz sands, limestones, and marly clays of the undifferentiated Cretaceous limestones, the sand and clay of the Woodbine Formation, the marl and clay of the Pecan Gap Chalk, the Navarro Group and Marlbrook Marl, the Tertiary

TABLE 3. GEOLOGIC UNITS AND SOIL RADON IN THE TEXAS COASTAL PLAIN (continued)

Geologic Unit	Lithology	# S	# N	Min.	Max
Cook Mountain Fm. (Ecm)	Clay, gypsiferous, silty, carbonaceous, glauconitic; and sandstone, glauconitic, very fine grained	2	0	—	—
Yegua Fm. (Ey)	Sandstone, glauconitic, fine grained, silty; and clay, carbonaceous	3	1	144	513
Caddell Fm. (Eca)	Clay, bentonitic, fossiliferous; local sandstone, glauconitic; and siltstone	2	0	10	64
Wellborn Fm. (Ewb)	Sandstone, quartz, fine to medium grained; and clay, carbonaceous	1	0	245	245
Whitsett Fm and Manning Fm (EOwh) (Em)	Sandstone, quartz; interbedded with clay, bentonitic	3	1	1,176	1,599
Catahoula Fm. (Mc)	Sandstone, quartz; interbedded with clay, bentonitic	0	0	—	—
Oakville Sandstone (Mo)	Sandstone, calcareous, medium grained; and clay, carbonaceous, reworked fossils	2	0	122	1,764
Fleming Fm. (Mf)	Sandstone, calcareous, medium grained; and clay, carbonaceous, reworked fossils	10	8	598	1,694
Willis Fm. (Qw)	Gravel, sand, and silt, clayey, limonitic	7	5	372	914
Lissie Fm. (Ql)	Sand, silt, clay, locally calcareous; minor gravel, limonitic, concretions	3	2	369	369
Quaternary Alluvium (Qal)	Floodplain deposits, mostly clay and silt; gravel and reworked terrace deposits.	5	2	617	908

Note: Letters in parentheses are the geologic map unit abbreviations taken from the maps of the Dallas, Sequin, and Abilene 2° sheets (University of Texas at Austin, 1972a,b, 1974) and are also used in Figure 1. K indicates Cretaceous; E, Eocene; M, Miocene; Q, Quaternary. Geologic descriptions were also derived from the Dallas, Sequin, and Abilene 2° sheets and our own field work. Units are listed from oldest to youngest.

Queen City Sand, the carbonaceous sands of Oakville sandstone and the Fleming and Wellborn Formations. Geologic units lowest in soil radon and eU include the quartz and glauconitic sands and clays of the Tertiary Wilcox, Weches, Yegua, Willis, and Cadell Formations, Tertiary Sand, the quartz sands of the Catahoula Formation, the Quaternary Willis Formation, the Lissie Formation, and river alluvium. Equivalent uranium was measured at only 23 sites in Texas; these data are plotted with soil radon in Figure 7. Soil radon data for all sites measured in Texas are plotted by composition in Figure 8. Carbonaceous sediments, glauconitic sediments, and chalk have the highest soil radon measurements. Limestones and quartzitic sediments yielded the lowest soil radon measurements.

DISCUSSION

Variation in radon concentrations due to sediment composition

Figure 9 shows the overall frequency distribution of soil radon concentrations in the Coastal Plain of all three states. The majority of the soils sampled were impermeable or had relatively low radon concentrations (<1,000 pCi/L). The highest concentrations of soil radon were measured in glauconitic, phosphatic, marly, and carbonaceous sediments. Recent geochemical work on the radioactivity of glauconite (Gundersen and Schumann, 1989) shows that uranium is concentrated throughout the glau-

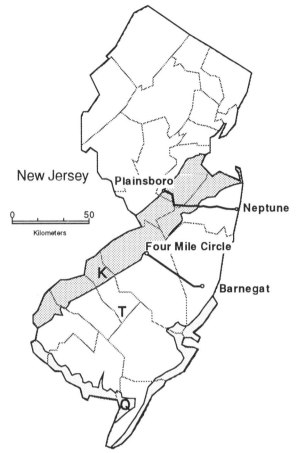

Figure 2. Locations of soil-gas sampling traverses in New Jersey.

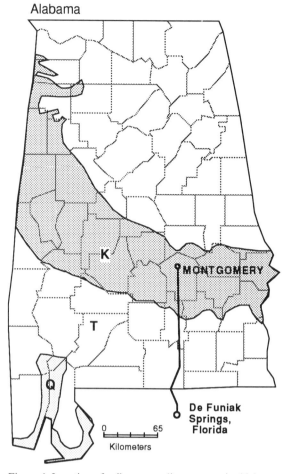

Figure 4. Location of soil-gas sampling traverse in Alabama.

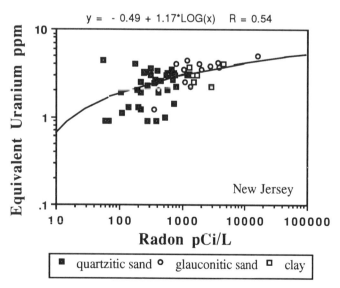

Figure 3. Plot of equivalent uranium vs. soil-gas radon by sediment composition for the New Jersey Coastal Plain.

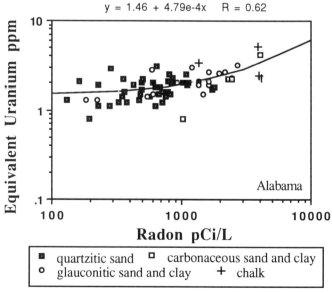

Figure 5. Plot of equivalent uranium vs. soil-gas radon by sediment composition for the Alabama Coastal Plain.

conite grain, especially near the grain surface and on cleavage faces, making it a concentrated and highly emanating source of radon. Phosphate is an effective scavenger of uranium, and phosphatic deposits tend to have high uranium concentrations, often higher than glauconitic sediments. The occurrence of fossils, especially those high in phosphate, such as shark teeth and whale bone, also correlates with moderate to high soil radon concentrations. A recent study of radon in soils from Coastal Plain sediments in Virginia, Berquist and others (1990) found uranium concentrated as high as 1,350 ppm in fossilized bone of the Yorktown Formation; P_2O_5 was 32% (C. R. Berquist, Jr., personal communication). They also found that the average radon was generally 1,000 pCi/L or less, the Yorktown sands having the highest average radon concentration in soil of 959 pCi/L. Organic material is widely recognized as a strong fixing agent for uranium and radium, humic materials in soils are particularly significant (Shanbhag and Choppin, 1981). Organic-rich sedimentary rocks such as black shales (Bell, 1978) are widely recognized as containing elevated to commercial amounts of uranium. From this study, carbonaceous sands and clays tended to produce the highest soil radon and have higher surface radioactivity, possibly due to their higher organic and uranium concentrations.

The correlation between soil radon and equivalent uranium is positive for all three states but is statistically significant only in Alabama. Deviation from a linear correlation for soil radon and equivalent uranium may be due to the effect of differences in permeability, radon emanation coefficient, and possible leaching of uranium from the uppermost soil horizon. Soils with high permeability tend to have higher concentrations of radon in the soil gas per amount of uranium in the soil; the gas is more accessible to measurement. Where the uranium and radium are sited in the soil is also important. In soils where uranium and radium are locked in mineral grains, the emanation of radon from that soil will be lower than in soils where uranium and radium are sited on grain surfaces or adsorbed onto clay, organic material, and metal-oxide surfaces. Leaching of radionuclides from the first foot of soil is common but dependent on soil and ground water chemistry (Durrance, 1986).

Variation in radon concentrations due to permeability

Data from the 64 sites measured using the MKII technique provide the opportunity to compare permeability directly with radon. The line graph of soil radon measured with the MKII vs. permeability (Fig. 10) indicates an optimum permeability for the higher radon concentrations. It is probable that, at higher permeabilities, atmospheric dilution lowers radon concentration, and at very low permeabilities radon transport is limited and therefore the source is limited. Permeability also varied greatly with soil texture (Fig. 11), suggesting that grain size is a poor indicator of soil permeability. Note that the clay category, which includes all the clays and chalks with distinctly high soil radon, has a number of permeability measurements in the 10^{-10} through 10^{-8} cm^2 range. This may also accentuate the number of high soil radon measurements in that permeability range.

Figure 6. Locations of soil-gas sampling traverses in Texas.

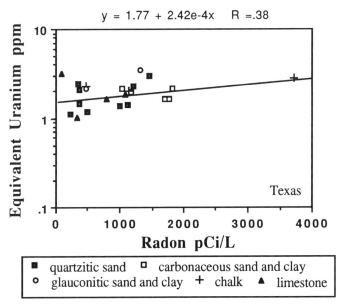

Figure 7. Plot of equivalent uranium vs. soil-gas radon by sediment composition for the Texas Coastal Plain.

Variation in radon concentrations due to technique and soil texture

Figure 12 shows a comparison between radon concentrations measured by the MKII and the RGS techniques. In well-sorted, medium- to coarse-grained sands, the two gas sampling techniques yield similar radon concentrations, usually within 10 percent of each other (Gundersen, 1992). This is likely due to the rigidity of the permeability, which is not altered greatly at the depth of 1 m by the pressure exerted on it by air flow in the MKII during sampling.

In poorly sorted loam, the RGS obtains the same or higher concentrations than the MKII. Higher concentrations measured using the RGS may possibly be due to collapse of the soil structure caused by the pressure exerted on the soil by the MKII. At several of the sampling sites with poorly sorted soils, a progressive increase in resistance to suction on the soil, as indicated on the pressure and flow meters of the MKII, was noted during sampling. In two cases, a soil-gas sample could not be obtained after

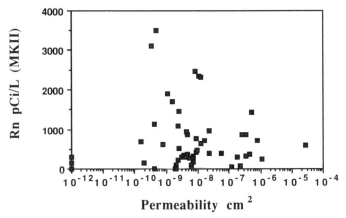

Figure 10. Plot of soil-gas radon vs. permeability measured with the MKII probe. Fourteen data points overlap at the origin.

the permeability measurement had been made because of apparent soil collapse.

In clay soils, several of the radon concentrations measured using the MKII were higher than the RGS. In clay samples, low permeability either prevents gas sampling or, as in dry clays, fractures create moderate to high permeability and easily obtained soil gas samples. It is not apparent why the MKII obtained higher results under these conditions than the RGS; however, it may possibly be related to the larger volume sampled and, in turn, to the larger emanating surface available to the MKII during measurement.

Moisture produces secondary effects on sampling and can determine whether a sample can be obtained at all. Under high moisture and saturated conditions, the MKII did not obtain a radon sample whereas the RGS obtained a sample 50% of the time. The RGS technique samples a much smaller volume and exerts less pressure on the soil, which accounts for its higher success rate in moist conditions. Several field experiments showed that under wet conditions, withdrawing more than 20 cm^3 of soil gas soon exhausted the available soil air and the attempt to draw a second or third sample yielded water in the soil probe.

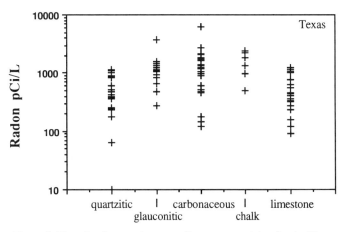

Figure 8. Plot of soil-gas radon vs. sediment composition for the Texas Coastal Plain.

Soil radon and indoor radon concentrations

Recent indoor radon surveys, conducted by the states of Alabama and Texas in cooperation with the Environmental Protection Agency, and by the state of New Jersey, revealed that the average indoor radon concentration is generally low (Muessig, 1989; Peake and Gundersen, 1989) in the Coastal Plain, even in the higher potential geologic units. The average indoor radon concentration over quartz sands and clays is less than 2 pCi/L. The average indoor radon concentration over phosphatic, carbonaceous, and glauconitic sediments is variable but is usually higher than 2 pCi/L. These indoor radon averages are lower than expected for comparable amounts of soil radon in other geologic

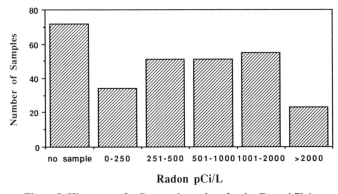

Figure 9. Histogram of soil-gas radon values for the Coastal Plain.

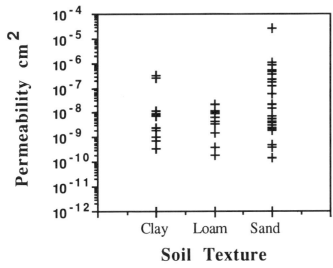

Figure 11. Plot of permeability vs. soil texture. Five samples in clay, seven samples in loam, and three samples in sand overlap at 10^{-12} cm^2 permeability.

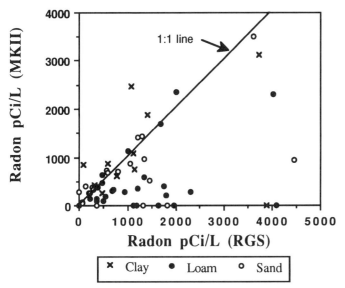

Figure 12. Plot of soil-gas radon measured with the MKII probe vs. soil-gas radon measured with the RGS probe, separated by soil texture.

terranes. For example, in the soils developed on the crystalline rocks of the Piedmont in the Appalachian Mountains, soil radon concentrations greater than 1,000 pCi/L may produce indoor radon concentrations of 10 pCi/L or more in homes with basements (Gundersen, 1989). The difference in the two terranes is most likely due to the architectural and lifestyle differences inherent in the populations of the two regions. In the Coastal Plain, houses tend to be slab on grade, be built on piers, or have crawl spaces rather than basements, whereas in the Piedmont, especially the northern states, homes tend to have basements.

ACKNOWLEDGMENTS

We thank C. Berquist, B. Hand, D. Owen, J. Smoot, and R. Schumann for their reviews of this manuscript, and G. Latzke, L. Hauser, and C. Wiggs for their participation in the field work. The research described in this work was funded in part by the U.S. Environmental Protection Agency.

REFERENCES CITED

Bell, R. Y., 1978, Uranium in black shales—A review, *in* Kimberley, M. M., ed., Uranium deposits, their mineralogy and origin: Mineralogical Association of Canada Short Course Handbook 3.

Berquist, C. R., Jr., Cooper, J. M., and Goodwin, B. K., 1990, Radon from Coastal Plain sediments, Virginia: Preliminary results: Geological Society of America Abstracts with Programs, v. 22, no. 2, p. 4–5.

Brooks, K. K., 1981, Geologic map of Florida: Gainsville, Florida Cooperative Extension Service, University of Florida, Institute of Food and Agricultural Sciences, scale 1:500,000.

Cohen, B. L., 1990, Surveys of radon levels in homes by University of Pittsburgh Radon Project, *in* Proceedings of the 1990 International Symposium on Radon and Radon Reduction Technology, Vol. III: Preprints: U.S. Environmental Protection Agency Report EPA/600/9-90/005c, Paper IV-3, 17 p.

Cohen, B. L., and Gromicko, N., 1988, Variation of radon levels in U.S. homes with various factors: Journal of the Air Pollution Control Association, v. 38, p. 129–134.

Copeland, C. W., Jr., 1988, Geologic map of Alabama: Alabama Geological Survey Special Map 220, 4 sheets, scale 1:250,000.

Durrance, E. M., 1986, Radioactivity and geology: Chichester, England: Ellis Horwood, 441 p.

Duval, J. S., Jones, W. J., Riggle, F. R., and Pitkin, J. A., 1989, Equivalent uranium map of the conterminous United States: U.S. Geological Survey Open-File Report 89-478, 10 p.

Dziuban, J. A., Clifford, M. A., White, S. B., Bergstein, J. W., and Alexander, B. V., 1990, Residential radon survey of twenty-three states, *in* Proceedings of the 1990 International Symposium on Radon and Radon Reduction Technology, Vol. III: Preprints: U.S. Environmental Protection Agency Report EPA/600/9-90/005c, Paper IV-2, 17 p.

Gundersen, L.C.S., 1989, Predicting the occurrence of indoor radon: A geologic approach to a national problem [abs.]: EOS Transactions of the American Geophysical Union, v. 70, no. 15, p. 280.

Gundersen, L.C.S., and Schumann, R. R., 1989, The importance of metal-oxides in enhancing radon emanation from rocks and soils: Geological Society of America Abstracts with Programs, v. 21, n. 6, p. A145.

Gundersen, L.C.S., 1992, The effect of rock-type, grain size, sorting, permeability, and moisture on measurements of radon in soil: A comparison of two measurement techniques: Journal of Radioanalytical and Nuclear Chemistry (in press).

Muessig, K. W., 1989, Radon in New Jersey: Proceedings of the 1988 Symposium on Radon and Radon Reduction Technology, U.S. Environmental Protection Agency Report EPA/ORP-89-600A.

Nielson, K. K., Bollenbacher, M. K., Rogers, V. C., and Woodruff, G., 1989, Users Guide for the MKII Radon/Permeability Sampler: U.S. Environmental Protection Agency Report RAE-8719/1-2.

Owens, J. P., and Minard, J. P., 1975, Geologic map of the surficial deposits in the Trenton area, New Jersey and Pennsylvania: U.S. Geological Survey Miscellaneous Investigation Map I-884.

Peake, R. T., and Gundersen, L.C.S., 1989, The Coastal Plain of the eastern and southern United States—An area of low radon potential: Geological Society of America Abstracts with Programs, v. 21, no. 2, p. 58.

Reimer, G. M., 1991, Simple techniques for soil-gas and water sampling for radon analysis, *in* Gundersen, L.C.S., and Wanty, R. B., eds., Field studies of radon in rocks, soils, and water: U.S. Geological Survey Bulletin 1971, p. 19–22.

Ronca-Battista, M., Moon, M., Bergsten, J., White, S. B., Holt, N., and Alexander, B., 1988, Radon-222 concentrations in the United States—Results of sample surveys in five states: Radiation Protection Dosimetry, v. 24, p. 307–312.

Schanbhag, P. M., and Choppin, G. R., 1981, Binding of uranyl by humic acid: Journal of Inorganic Nuclear Chemistry, v. 43, no. 12, p. 3369–3372.

Schumann, R. R., Peake, R. T., Schmidt, K. M., and Owen, D. E., 1990, Correlations of soil-gas and indoor radon with geology in glacially derived soils of the northern Great Plains, *in* Proceedings of the 1990 EPA International Symposium on Radon and Radon Reduction Technology, Volume III: Preprints: U.S. Environmental Protection Agency Report EPA/600/9-90/005c, Paper VI-3, 14 p.

University of Texas at Austin, 1972a, Geologic atlas of Texas, Abilene Sheet: Bureau of Economic Geology, University of Texas at Austin, scale 1:250,000.

—— , 1972b, Geologic atlas of Texas, Dallas Sheet: Bureau of Economic Geology, University of Texas at Austin, scale 1:250,000.

—— , 1974, Geologic atlas of Texas, Sequin Sheet: Bureau of Economic Geology, University of Texas at Austin, scale 1:250,000.

Zapecza, O. S., 1989, Hydrogeologic framework of the New Jersey Coastal Plain: U.S. Geological Survey Professional Paper 1404-B, 49 p.

MANUSCRIPT ACCEPTED BY THE SOCIETY APRIL 6, 1992

Effects of weather and soil characteristics on temporal variations in soil-gas radon concentrations

R. Randall Schumann, Douglass E. Owen, and Sigrid Asher-Bolinder
U.S. Geological Survey, Box 25046, Federal Center, MS-939, Denver, Colorado 80225

ABSTRACT

Concentrations of radon-222 in soil gas measured over about 1 yr at a monitoring site in Denver, Colorado, vary by as much as an order of magnitude seasonally and as much as severalfold in response to changes in weather. The primary weather factors that influence soil-gas radon concentrations are precipitation and barometric pressure. Soil characteristics are important in determining the magnitude and extent of the soil's response to weather changes. The soil at the study site is clay rich and develops desiccation cracks upon drying that increase the soil's permeability and enhance gas transport and removal of radon from the soil. A capping effect caused by frozen or unfrozen soil moisture is a primary mechanism for preventing radon loss to the atmosphere.

INTRODUCTION

Although geologic and soil characteristics are primarily responsible for controlling the amount of radon generated by soils, soil-gas radon concentrations vary under the influence of climatic and meteorologic factors. An understanding of seasonal and weather-related variations in soil-gas radon concentrations is essential for accurate site evaluations and can aid in planning indoor radon testing programs and mitigation schemes. Meteorologic factors affect the emanation, migration, and concentration of radon. A general discussion of radon generation and transport processes, and how they are affected by meteorologic factors, provides a background for the discussion of empirical data from this study.

The amount of radon generated by a rock or soil that is available to the pore space is termed the emanation coefficient of the soil. Radon emanation rates are highest when soil moisture is between 15 and 20% by weight (Damkjaer and Korsbeck, 1985; Lindmark and Rosen, 1985; Stranden and others, 1984). At these soil moisture levels, pore water exists as thin coatings on soil grains. The water absorbs some of the recoil energy of radon atoms as they escape, preventing the atoms from burying themselves in adjacent soil grains and thus increasing the chance that the recoil path will terminate in a pore space (Tanner, 1964, 1980). At higher levels of soil moisture, radon mobility is affected because the thicker liquid coating on soil grains traps the radon atoms in the pore liquid. They may then move by diffusion through the liquid or into the gas phase, where they can then move by diffusion, convective flow, or a combination of both, provided the soil is not saturated and the gas-filled pores are interconnected.

Radon transport in soils occurs by two processes, diffusion and convective flow. Diffusion is the process by which radon atoms move through the pore fluids (gases and/or liquids) in response to a concentration gradient, as described by Fick's Law. Convective transport occurs when the pore fluids move through the soil pores driven by a pressure gradient as described by Darcy's Law, carrying the radon atoms along with them. Diffusion is the dominant radon transport process in soils of lower permeability (generally less than 10^{-7} cm^2), whereas convective transport processes tend to dominate in more permeable soils (generally greater than 10^{-7} cm^2) (Sextro and others, 1987).

Meteorologic conditions have a marked effect on radon transport in soils. In this discussion, we assume an approximately direct correlation between wet and dry weather periods and soil moisture conditions. This assumption is an oversimplification if individual precipitation events are examined because it disregards the importance of prior soil moisture conditions. The relationship between precipitation and soil moisture is also affected by seasonal differences in evapotranspiration rate, which influence how

---

Schumann, R. R., Owen, D. E., and Asher-Bolinder, S., 1992, Effects of weather and soil characteristics on temporal variations in soil-gas radon concentrations, *in* Gates, A. E., and Gundersen, L.C.S., eds., Geologic Controls on Radon: Boulder, Colorado, Geological Society of America Special Paper 271.

much of the precipitation that falls on the ground surface is available to infiltrate the soil. In general, however, this relationship is valid if evaluated in a seasonal context.

Soil moisture affects both radon emanation and transport. As discussed previously, radon emanation is enhanced at low to moderate soil moisture levels. Radon transport is generally inhibited by soil moisture because, as soil moisture content increases, water can block soil pores, reducing both the radon diffusion coefficient and the gas permeability of the soil. Radon atoms can move by diffusion through water, but whereas a radon atom can travel 1 to 2 m by diffusion through dry soil during its mean lifespan, it may migrate only 1 to 2 cm in saturated soil during the same time period (Søgaard-Hansen and Damkjaer, 1987). In finer grained soils, especially those with high clay content, less moisture is necessary to inhibit radon transport because of three factors: (1) the pore spaces are smaller and thus can be more easily blocked by soil water; (2) interlayer water molecules are electrostatically bound to the clay particles, so clay-rich soils hold moisture longer and tend to dry out more slowly; and (3) expandable clays swell with the addition of moisture, closing pore spaces and cracks in the soil more readily than in a coarser grained soil.

Changes in the partitioning of radon between the gas and water phases in soil pores may cause as much as fivefold variations in soil-gas radon content. The partitioning is controlled by the amount of soil moisture present and by temperature (Rose and others, 1990; Washington and Rose, 1990). Spatial differences in soil moisture regimes, which generally follow climatic zones, may significantly influence seasonal soil-gas radon concentrations. For example, Rose and others (1990) suggested that, in cooler, wetter climates, summer soil radon concentrations will be higher than winter concentrations, whereas winter soil radon concentrations may typically exceed summer concentrations in soils of drier climates.

Capping is a moisture-related effect that tends to increase measured soil-gas radon concentrations. A capping effect occurs when the uppermost soil layers become saturated or the moisture in them is frozen, inhibiting the release of radon to the atmosphere and allowing radon to accumulate beneath the capping layer. The capping layer forms a barrier between the soil and the atmosphere, suppressing barometric, thermal, and wind effects. Rainfall can produce an effective moisture cap by filling pores in the near-surface soil layers (Kraner and others, 1964; Kovach, 1945; Schery and others, 1984; Schumann and Owen, 1988). Freezing of the moisture in the uppermost soil layers also appears to be a common and efficient capping mechanism (Hesselbom, 1985; Kovach, 1945; Lindmark and Rosen, 1985). The capping effect may be enhanced during spring and fall, when the diurnal freeze-thaw cycle allows moisture to infiltrate the near-surface soil layers during the day and subsequently to freeze at night. Capping may occur more readily in smectitic soils because the surface layers swell shut, blocking radon exhalation and slowing further infiltration of moisture. This effect may be quite dramatic in soils with clayey B horizons that act as the capping layer (Schumann and Owen, 1988).

Barometric pressure changes have been found to cause significant temporary changes in measured soil-gas radon concentrations. Falling pressure tends to draw soil gas out of the ground, increasing the radon flux across the soil/air interface, which may increase the radon concentration in the near-surface layers. Conversely, high or increasing barometric pressure forces atmospheric air into the soil, diluting the near-surface soil gas and driving radon deeper into the soil (Hesselbom, 1985; Kovach, 1945; Kraner and others, 1964; Lindmark and Rosen, 1985). Bakulin (1971, p. 125) found that a decrease in pressure causes an increase in radon exhalation "proportional to the square of the pressure drop rate, [and to the] square of the gas permeable soil layer depth, and [the exhalation rate] increases linearly with time." Clements and Wilkening (1974) noted that pressure changes of 1 to 2 percent associated with the passage of weather fronts could produce changes of 20 to 60 percent in the radon flux, depending on the rate of change of pressure and its duration. Wind turbulence and the Bernoulli effect imparted by wind blowing across an irregular soil surface can draw soil gas upward from depth in a manner similar to that of decreasing barometric pressure (Hesselbom, 1985; Kovach, 1945; Pearson and Jones, 1966).

Some authors (for example, Kovach, 1945; Lindmark and Rosen, 1985) have suggested that temperature has little or no effect on soil gas radon content. However, Ball and others (1983) found that soil-gas radon concentrations correlate with changes in soil temperature and, to a lesser extent, with air temperature changes. Kovach (1945) reported higher radon exhalation during temperature lows. Klusman and Jaacks (1987) observed negative correlations between both soil and air temperature and radon concentrations, and suggested that temperature gradients within the soil, or between the soil and air, can induce convective soil-gas transport. However, a theoretical analysis based on energy balance equations led Schery and Petschek (1983) to conclude that thermal gradients within soils are insufficient to induce measurable convective transport of radon.

A temperature effect on radon exhalation has also been noted. In one experiment, radon exhalation rates in soil and shale samples increased by 50 to 200% in response to increasing the temperature from 5 to 22°C (Stranden and others, 1984), whereas another experiment yielded an approximately 10% increase in radon activity when granite samples were heated from −20 to 22°C (Barretto, 1975).

SITE DESCRIPTION

The Denver Federal Center (DFC) soil-gas monitoring site, located in Denver, Colorado (Fig. 1), is situated on flat-lying, unirrigated ground at least 100 m from the nearest building, major highway, or other man-made structure. The climate of the area is semiarid, with mean annual precipitation of 36 to 40 cm,

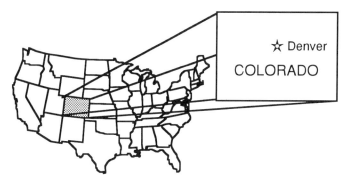

Figure 1. Location of study site.

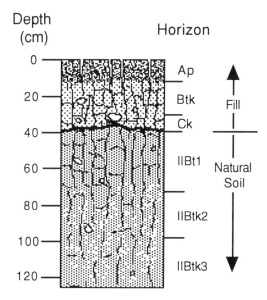

Figure 2. Soil profile at the DFC study site. Ap = human-disturbed surface mineral horizon (clayey silt loam) with significant organic matter content; Btk = subsurface silty clay loam horizon containing illuvial clay and calcium carbonate ($CaCO_3$); Ck = subsurface horizon consisting primarily of sandy loam, with $CaCO_3$ coatings on grains and in pore spaces; IIBtl = clay to silty clay B horizon of buried soil layer; IIBtk2 = clay to silty clay similar to IIBtl, but with significant accumulations of $CaCO_3$ coatings on grains and in pore spaces; IIBtk3 = silty clay subsurface horizon similar to above, but with more intense $CaCO_3$ development. The soil has a strong prismatic structure throughout the profile described above.

about 70 percent of which falls in the spring and summer months. Spring weather is characterized by several days of cool temperatures and rain or snowfall alternating with periods of mild, dry weather, whereas summer is generally warm and dry except for occasional late-day, short-duration thunderstorms. Winters are cool and relatively dry, with daytime temperatures that fluctuate above and below freezing throughout the season.

The soil underlying the DFC site is a clay to silty clay derived from alluvium, mudstone, and shale, but it also contains pebbles and cobbles. Much of the clay is smectitic (swelling). Below about 70-cm depth the soil contains $CaCO_3$ in grain coatings and small concretions (Fig. 2). The soil is overlain by approximately 30 cm of locally derived fill similar in appearance and soil properties to the underlying natural soil. The B and C horizons of the fill soil also contain $CaCO_3$ coatings on soil grains and rock fragments. Because of its high clay content and shrink-swell potential, both the fill and natural soil develop a system of desiccation cracks that extend to a depth of at least 1.2 m (the maximum depth to which a trench was excavated to examine the soil profile) and impart a prismatic structure to the soil (Fig. 2). The crack system has both vertical and horizontal components, but the vertical cracks are more extensive and better developed. Because it is composed primarily of clay, the soil has a low intergranular permeability, but the crack system imparts a moderate gas permeability (approximately 10^{-9} cm^2 at 100 cm; R. T. Peake, oral communication, 1987) to the soil under dry conditions. Cracks were present from May to October, except for short periods following storms. The average radium-226 content of the soil is approximately 0.8 pCi/g, ranging from about 0.6 to 1.0 pCi/g in six samples collected from this profile.

METHODS

Concentrations of radon in soil gas were monitored from March 1987 to April 1988 at the DFC study site. The monitoring site instrumentation consists of three stainless steel soil gas probes, approximately 8 mm in outside diameter, inserted to depths of 50, 75, and 100 cm. A surface radon collection chamber consisting of a polystyrene box measuring approximately 30 × 45 × 10 cm was constructed to sample radon exhaled from the soil surface. The box was installed in December 1987 by inverting the box on the soil surface and packing soil around its edges to seal it from the atmosphere. Soil gas samples were extracted from each probe and from the surface radon collection chamber with syringes and injected into the sampling cell of a portable alpha scintillation-type radon detector. Accuracy of the radon concentration measurements is approximately ±10%. Samples were collected between 10 a.m. and 2 p.m. each day to minimize diurnal variability. Additional information on the probe design and methods used is given by Reimer (1990). Air and soil temperatures were monitored at the ground surface and at depths of 10 and 60 cm. Tensiometers and gypsum block soil-moisture monitors were installed at the site, but the data from these devices were considered unreliable and were not used. Precipitation data were used instead as a rough indicator of soil moisture conditions. Soil-gas radon concentrations were compared with weather data from a nearby permanent weather station operated by the U.S. Bureau of Reclamation and with soil temperature data collected at the site. One soil-gas radon measurement was made at each depth each weekday; soil temperature was measured once each weekday at the same time; all other weather data were collected hourly.

WEATHER AND SOIL-GAS RADON AT THE DFC SITE

Soil-gas radon concentrations at the DFC site exhibit a marked seasonality, with an order-of-magnitude difference between seasonal soil-gas radon maxima and minima (Fig. 3). Radon concentrations were highest in the late winter and early spring, a period characterized by relatively wet, unstable weather, and were lowest in the fall, a season with typically dry, stable weather. Soil-gas radon concentrations generally increase with depth. Radon concentrations at all sampled depths appear to react similarly to seasonal and shorter term meteorologic changes, which indicates that sufficient vertical permeability exists for atmospheric influences to affect soil gases at depths of 1 m or more during most of the year.

During the period of late January to mid-March 1988, radon concentrations at 100-cm depth could not be measured because apparent excessive soil moisture conditions at that depth prevented gas samples from being collected. In addition, the seemingly anomalously low soil-gas radon values at 100 cm recorded in late March and early April 1987 may be due to low gas permeability caused by wet soil conditions at that depth. That wet soil conditions existed at that time is suggested by the wet soil that existed prior to the start of the study, compounded by regularly occurring precipitation throughout March and early April (Fig. 4). Becasue we did not have 50- and 75-cm soil gas probes or soil moisture monitors installed at that time, however, we cannot confirm that the soil column was uniformly saturated.

Seasonality is reflected in most of the weather variables. Precipitation events were of greater intensity, duration, and frequency from March through early July, 1987, followed by a 6-mo dry period during which there were only four rainy periods (Fig. 4). Barometric pressure fluctuations (Fig. 4) also follow seasonal trends. Larger variations in barometric pressure associated with frontal weather systems during the period of March to June 1987, were followed by a period of relatively stable weather with less pronounced barometric pressure variations from June through mid-December, except for two storms in late October and early November. Air and soil temperatures follow seasonal

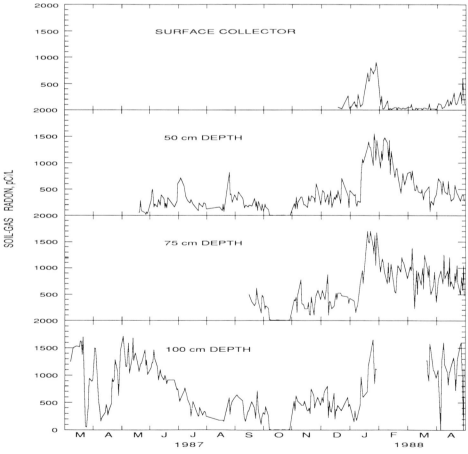

Figure 3. Radon concentrations in the surface radon collector and in soil gas at 50-, 75-, and 100-cm depths from March 1987 through April 1988 (note: 1 pCi/L = 37 Bq/m^3).

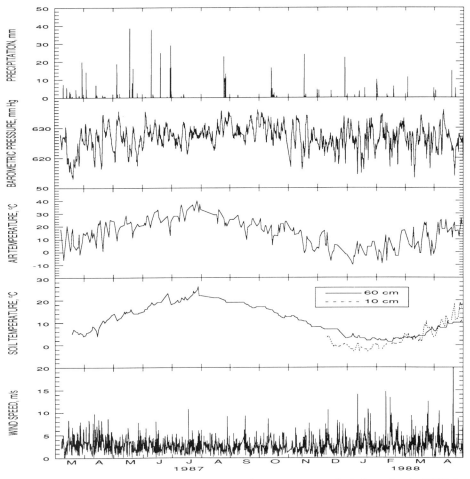

Figure 4. Precipitation, barometric pressure, air and soil temperatures, and wind speed at the DFC site for the period from March 1987 through April 1988.

trends as well (Fig. 4). Late winter to early spring is typically a period with increased wind velocities (Fig. 4), but wind does not appear to correlate with changes in soil-gas radon concentrations at the DFC site.

Longer term seasonal trends are those in which higher radon values (hundreds to thousands of picocuries/liter) are associated with unstable, wet weather periods, and lower radon values (tens to low hundreds of picocuries/liter) are associated with dry, stable weather periods. In contrast, shorter term variations in soil-gas radon concentrations are caused by changes in weather associated with discrete storms. There is a relatively good positive correlation between precipitation and radon in soil gas at all depths (an example is shown in Fig. 5). One exception is the precipitation event in mid-October that had virtually no effect on soil-gas radon concentrations, which illustrates the importance of antecedent conditions in influencing the extent of the soil's response to weather changes. Because the October storm was preceded by a long period of dry weather, the soil was apparently so dry that the approximately 25 mm of precipitation were insufficient to seal soil cracks or the soil surface at the site, and thus had

little or no effect on soil-gas radon concentrations. In contrast, each rainfall in April, May, and early June caused soil-gas radon concentrations at the 100 cm-depth to increase by several hundred picocuries/liter because soil moisture levels were already high, so that less moisture was required to induce capping and close surface cracks, and because rainfall amounts from spring storms were generally greater than those of summer and fall storms. It appears that capping is the most important weather-related effect on the DFC site. The intensity of this effect is due largely to soil characteristics, particularly the presence of significant amounts of swelling clay and desiccation cracks.

Barometric pressure correlates negatively with soil-gas radon concentrations over time periods of a few days, in that falling pressure is associated with increasing concentrations and vice versa (Fig. 5). As in the case of precipitation, the smaller peaks and troughs on the radon curves coincide with changes in barometric pressure. The correlations appear to be better during the "stormy" seasons (for example, during May through July) than during the "stable weather" seasons, possibly because the magnitude and rate of pressure change are more important than

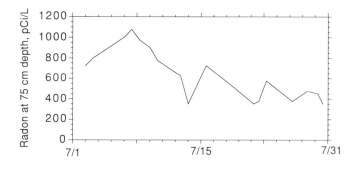
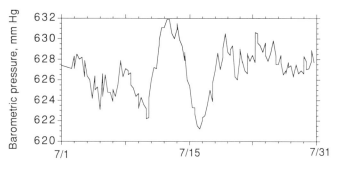
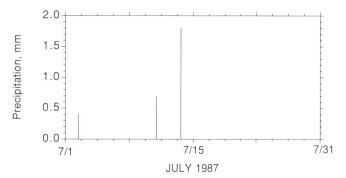

Figure 5. Soil-gas radon, barometric pressure, and precipitation at the DFC study site, July 1987.

the actual value of the barometric pressure. For example, a change in barometric pressure from 630 to 625 mm Hg would be expected to cause an increase in soil-gas radon concentration similar to that of a change from 625 to 620 mm Hg because the amount of pressure change is the same in both cases.

A change in barometric pressure produces a pressure gradient between the atmosphere and soil that induces vertical soil gas transport. The strength of this effect is determined by the magnitude of the pressure gradient. Thus a 10-mm change from 630 to 620 mm Hg should have a more pronounced effect than a 5-mm change from 625 to 620 mm Hg. If the magnitude of the barometric pressure change is relatively small, or the change is gradual, the gas pressure in the soil pores should be able to more easily equilibrate with the barometric pressure of the air, so pressure-gradient induced soil gas transport would be less likely to occur or its effect may be diminished. In addition, pressure-induced transport can occur only if the soil is permeable enough to allow vertical convective transport to occur. In theory, because the major permeability in the DFC soil is due to desiccation cracks, this phenomenon is more likely to occur when the soil is dry enough for a network of desiccation cracks to exist, although this is not clearly confirmed by the data.

The effect of temperature on soil-gas radon concentrations appears to be minor compared to those of precipitation and barometric pressure. Air-temperature lows appear to correspond with soil-gas radon highs, but there is no discernable correlation between soil temperature at a depth of 60 cm and soil-gas radon concentrations (Figs. 3, 4). Some researchers suggest that temperature gradients between the soil and air could induce thermal convection that would cause soil gas to flow (Kovach, 1945; Klusman and Jaacks, 1987). Soil permeability should be important in determining whether such convective transport is possible in a given soil. At the DFC site, soil permeability is low except when enhanced by cracking. Because of low soil permeability and in light of calculations by Schery and Petschek (1983) showing that measurable thermally induced convection would not be significant in "normal" soils, we conclude that it is unlikely that thermally induced convection is a major driving force at this site.

Another temperature-related phenomenon may play a significant role in radon accumulation during winter months— capping effects due to freezing of water in the uppermost soil layers. This is likely an important factor in producing elevated indoor radon levels during winter months in many areas. Moisture capping occurs when precipitation infiltrates the uppermost soil layers, filling pores and causing clays to swell and cracks to close. Because of its low intergranular permeability, percolation through clayey soil is slow, so that pores and cracks deeper in the soil may remain open for a considerable amount of time after the surface cracks have swelled shut. Because it cannot escape to the surface, radon may accumulate to elevated levels beneath the capping layer. An example of the capping effect at the DFC site occurred in early January 1988, when radon concentrations at all depths, including surface radon exhalation, increased from a few hundred to more than 1,000 pCi/L (Fig. 3). These increases coincided with minor storms on several occasions (compare with precipitation and barometric pressure curves, Fig. 4), but more importantly, with the period during which the near-surface soil temperature (10-cm depth soil temperature curve, Fig. 4) dropped and remained below freezing.

Although the soil-gas radon concentrations increased and remained elevated for several months beginning in early January 1988, the surface radon exhalation decreased after about 1 mo, which corresponded with an increase in soil temperature at 10-cm depth from below to above freezing (Fig. 4). A proposed explanation is that the surface radon collection chamber initially prevented moisture from infiltrating the soil directly beneath it. When the surrounding soil froze, soil gas was still able to escape into the collection chamber. Additional precipitation during January was unable to infiltrate the soil, and moisture in the uppermost soil layers could not move laterally because it was frozen. After the capping layer thawed, a liquid moisture cap remained. The soil-gas radon concentrations stayed elevated, but soil moisture was able to migrate laterally, saturating the soil beneath the surface radon collector and inhibiting radon escape, so radon levels in the chamber dropped. A moisture cap is more easily maintained during cooler months because evapotranspiration rates are lower, so the soil dries out more slowly. At the DFC site, the latter part of the winter corresponds roughly with the first part of the wetter season (spring and early summer), providing favorable conditions for moisture caps (frozen and unfrozen).

SUMMARY AND CONCLUSIONS

Radon concentrations in soil gas at the DFC site vary by as much as an order of magnitude between seasons, and by as much as 200 percent in response to day-to-day weather variations. Although the seasonality of these variations may be different in different areas, similar magnitudes of variations occur in other soils and climatic zones, one example of which is in central Pennsylvania (Washington and Rose, 1990).

Soil characteristics are important in determining the response of a particular soil to climatic and weather factors. Soil mineralogy, structure, grain size distribution, and sorting control the permeability, moisture infiltration rates, the presence and extent of cracks, swelling, and similar factors that influence the soil's ability to inhibit or enhance radon transport within the soil or to release soil gas to the atmosphere. In the case of the DFC site, the clayey texture and dry climate interact to produce the desiccation crack system that imparts significant gas permeability to an otherwise low-permeability soil. The swelling of expandable clays in response to the addition of a relatively small amount of moisture causes rapid changes in soil permeability, which may exaggerate their responses to moisture input in comparison with soils of different composition, even in similar climates.

Determining the influence of individual weather factors and their effects on radon migration and concentration in soils is complicated by the fact that several weather factors may change simultaneously. For example, storms are generally associated with precipitation, lower barometric pressure, lower temperatures, and wind. Comparing plots of these factors with soil-gas radon concentrations may lead to the conclusion that any or all of these factors cause soil-gas radon concentrations to increase. Because changes in all of the weather factors often occur together, the task of determining the relative influence of each individual factor is difficult. Unless the effect of each factor can be isolated in a controlled environment, it may not be possible to definitively determine the relative importance of each weather factor.

Seasonal variations are by far the most striking of those observed during this study. The combination of weather factors associated with different seasonal weather regimes produced a relatively consistent set of conditions for sustaining high or low soil-gas radon levels for extended periods of time. Whereas short-term weather variations caused measured soil-gas radon concentrations to vary by as much as a factor of 2, seasonal extremes in radon values differed by as much as a factor of 10.

The most important weather-related parameter affecting soil-gas radon concentrations is soil moisture, and by association, precipitation, as it provides the moisture. Nearly every precipitation event recorded during the study period evoked a noticeable change in radon concentrations in soil gas. Barometric pressure appears to be the next most important parameter. Inverse correlations between barometric pressure and soil-gas radon values were noted during periods when large-scale barometric pressure changes occurred without associated precipitation. The magnitudes of changes in radon concentration in response to barometric pressure changes alone were generally less than that of precipitation alone or of a combination of the two. The influences of temperature, wind, and other weather factors are not specifically determined but appear to be of lesser importance, because the more dominant factors tend to overshadow the effects of those factors with weaker influences.

The data presented here indicate that evaluations of building tracts or sites made on the basis of a single or a few soil-gas radon measurements could be misleading if they are made without considering the potential variability imposed by seasonal and day-to-day weather variations. Additional data from other soil types and climatic zones should be collected to provide a better basis for evaluating the validity of individual short-term soil-gas radon measurements for use in long-term predictions of radon potential. The basic concepts and relationships described in this paper should be applicable to virtually any area if careful investigation and interpretation of soil properties and seasonal climatic variations at each site are undertaken.

ACKNOWLEDGMENTS

We thank Sherry Agard, U.S. Geological Survey, for describing the soil profile at the DFC site. Reviews by J. Herring, J. Otton, H. Rector, and A. Tanner greatly improved the manuscript.

REFERENCES CITED

Bakulin, V. N., 1971, Dependence of radon exhalation and its concentration in the soil on meteorological conditions (abs): Chemical Abstracts, v. 74, no. 33637u, p. 125.

Ball, T. K., Nicholson, R. A., and Peachey, D., 1983, Effects of meteorological variables on certain soil gases used to detect buried ore deposits: Transactions Institution of Mining and Metallurgy, v. 92, p. B183–B190.

Barretto, P.M.C., 1975, Radon-222 emanation characteristics of rocks and minerals, in Radon in uranium mining: Panel proceedings, Vienna, IAEA-PL-565-1, p. 129–150.

Clements, W. E., and Wilkening, M. H., 1974, Atmospheric pressure effects on Rn-222 transport across the earth-air interface: Journal of Geophysical Research, v. 79, p. 5025–5029.

Damkjaer, A., and Korsbeck, U., 1985, Measurement of the emanation of radon-222 from Danish soils: The Science of the Total Environment, v. 45, p. 343–350.

Hesselbom, A., 1985, Radon in soil gas—A study of methods and instruments for determining radon concentrations in the ground: Sveriges Geologiska Undersokning, ser. C, no. 803, p. 1–58.

Klusman, R. W., and Jaacks, J. A., 1987, Environmental influences upon mercury, radon and helium concentrations in soil gases at a site near Denver, Colorado: Journal of Geochemical Exploration, v. 27, p. 259–280.

Kovach, E. M., 1945, Meteorological influences upon the radon content of soil gas: EOS, Transactions of the American Geophysical Union, v. 26, p. 241–248.

Kraner, H. W., Schroeder, G. L., and Evans, R. D., 1964, Measurements of the effects of atmospheric variables on radon-222 flux and soil gas concentrations, in Adams, J.A.S., and Lowder, W. M., eds., The natural radiation environment: Chicago, University of Chicago Press, p. 191–215.

Lindmark, A., and Rosen, B., 1985, Radon in soil gas—Exhalation tests and in situ measurements: The Science of the Total Environment, v. 45, p. 397–404.

Pearson, J. E., and Jones, G. E., 1966, Soil concentrations of "emanating radium-226" and the emanation of radon-222 from soils and plants: Tellus, v. 18, p. 655–661.

Reimer, G. M., 1990, Reconnaissance techniques for determining soil-gas radon concentrations: An example from Prince Georges County, Maryland: Geophysical Research Letters, v. 17, p. 809–812.

Rose, A. W., Ciolkosz, E. J., and Washington, J. W., 1990, Effects of regional and seasonal variations in soil moisture and temperature on soil gas radon, in Proceedings of the 1990 International Symposium on Radon and Radon Reduction Technology, Volume 3: Preprints: U.S. Environmental Protection Agency Report EPA/600/9-90/005c, Paper C-VI-5.

Schery, S. D., and Petschek, A. G., 1983, Exhalation of radon and thoron: The question of thermal gradients in soil: Earth and Planetary Science Letters, v. 64, p. 56–60.

Schery, S. D., Gaddert, D. H., and Wilkening, M. H., 1984, Factors affecting exhalation of radon from a gravelly sandy loam: Journal of Geophysical Research, v. 89, p. 7299–7309.

Schumann, R. R., and Owen, D. E., 1988, Relationships between geology, equivalent uranium concentration, and radon in soil gas, Fairfax County, Virginia: U.S. Geological Survey Open-File Report 88-18, 28 p.

Sextro, R. G., Moed, B. A., Nazaroff, W. W., Revzan, K. L., and Nero, A. V., 1987, Investigations of soil as a source of indoor radon, in Hopke, P. K., ed., Radon and its decay products: American Chemical Society Symposium Series 331, p. 10–29.

Søgard-Hansen, J., and Damkjaer, A., 1987, Determining ^{222}Rn diffusion lengths in soils and sediments: Health Physics, v. 53, p. 455–459.

Stranden, E., Kolstad, A. K., and Lind, B., 1984, Radon exhalation: Moisture and temperature dependence: Health Physics, v. 47, p. 480–484.

Tanner, A. B., 1964, Radon migration in the ground: A review, in Adams, J.A.S., and Lowder, W. M., eds., The natural radiation environment: Chicago, University of Chicago Press, p. 161–190.

——— , 1980, Radon migration in the ground: A supplementary review, in Gesell, T. F., and Lowder, W. M., eds., Natural radiation environment 3, Symposium Proceedings, Houston, Texas, v. 1, p. 5–56.

Washington, J. W., and Rose, A. W., 1990, Regional and temporal relations of radon in soil gas to soil temperature and moisture: Geophysical Research Letters, v. 17, p. 829–832.

MANUSCRIPT ACCEPTED BY THE SOCIETY APRIL 6, 1992

A theoretical model for the flux of radon from rock to ground water

Richard B. Wanty and Errol P. Lawrence*
U.S. Geological Survey, Box 25046, Federal Center, MS-916, Denver, Colorado 80225
Linda C. S. Gundersen
U.S. Geological Survey, Box 25046, Federal Center, MS-939, Denver, Colorado 80225

ABSTRACT

A model is derived to predict the abundance of ^{222}Rn in ground water in contact with a rock of known uranium content. The model assumes that secular equilibrium is attained in the rock-water system as a whole, but is independent of any microscopic geometric properties of the system. The key variables in the model are bulk properties such as porosity, uranium content of the rock, emanating efficiency, and rock density, all of which are measurable. Thus, the model is simplified by the averaging effects of a macroscopic view of the system. Although less rigorous than other models presented in the literature, it is more generally applicable to natural systems because it does not rely on microscopic properties of the system, which are impossible to quantify. Application of the model to crystalline aquifers in the eastern United States shows that bulk emanation rates of radon are generally less than about 30%.

INTRODUCTION

Recent discovery of high levels of ^{222}Rn in indoor air in homes has led to intensive investigations of the natural processes controlling levels of radon in soil gas and ground water (cf. Nazaroff and Nero, 1988; Gundersen and others, 1988; Wanty and Gundersen, 1988). It is uncertain whether radon in drinking water poses a direct health threat (Cross and others, 1985; Cothern, 1987; Crawford-Brown, 1990); the main concern over high levels of radon in ground water is that levels of radon in indoor air can be supported by radon derived from the water supply (cf. Prichard and Gesell, 1981; Nazaroff and others, 1988; Hess and others, 1990; Lawrence and others, 1992). Thus, radon in ground water is of concern to those using the ground water for domestic supply.

Numerous field studies have been conducted and many models for radon generation have been proposed. A brief review of these models follows; see also Semkow (1990) for a discussion of existing models. Most of the models are specifically tailored for a single field area or have been developed using microscopic properties such that application of the models to a natural system is difficult or impossible. Thus, a practical macroscopic theory for radon emanation from rocks to ground water has not yet been developed. Such a theory would allow prediction of levels of radon in ground water if the uranium concentration of the aquifer rock is known. This predictive capability would facilitate preliminary assessments of extensive geographic areas through the use of large sets of data for uranium in rocks from programs such as the National Uranium Resources Evaluation (NURE). This chapter presents a simple model for the transfer of radon from rock to ground water. The treatment here differs from previous models in that a more macroscopic approach is followed. Thus, the equations derived here are simplified by the averaging effects of examining a larger system.

One of the earliest papers to mathematically describe the phenomenon of radon emanation was that of Flügge and Zimens (1939). Those authors derived a set of equations to describe emanation of radon as a combination of recoil and diffusion processes, and showed that under most conditions the recoil fraction far outweighs the diffusion fraction. More recently, Fleischer (1983, 1988) and Semkow (1990) developed emanation models that described recoil release of radon into pores and considered

---

*Present address: Geraghty and Miller, Inc., 1099 18th Street, Suite 2100, Denver, Colorado 80202.

Wanty, R. B., Lawrence, E. P., and Gundersen, L.C.S., 1992, A theoretical model for the flux of radon from rock to ground water, *in* Gates, A. E., and Gundersen, L.C.S., eds., Geologic Controls on Radon: Boulder, Colorado, Geological Society of America Special Paper 271.

aspects of surface geometry and radium distribution in the solid. These models led to reasonable fits of experimental data, although they included many parameters that are impossible to measure practically, including specific surface area of the solid, surface roughness, and inhomogeneous ^{226}Ra distribution in the solid. Thus, although the models of Flügge and Zimens (1939), Fleischer (1983, 1988), and Semkow (1990) are satisfying in their rigorous treatment of surface characteristics, no real predictive capability is gained by them.

One interesting aspect of the models of Fleischer (1983, 1988) and Semkow (1990) is in the comparison of wet to dry systems. They found that, in most cases, wet systems had higher emanating efficiencies than dry ones, primarily due to decreased imbedding of recoiling ^{222}Rn atoms in surfaces across a pore. These studies are supported by experiments of Thamer and others (1981), who found emanation coefficients 1.5 to 4 times greater for wet uranium ore samples than for dry ones. In wet systems, fewer ^{222}Rn atoms recoil across a pore into the far wall because the recoil distance in water is about 700 times less than that in air (cf. Semkow, 1990). In addition to the stopping power of water, Fleishcer (1983) credited the leaching effects of water on increasing radon emanation.

A series of recent papers by Hammond and others (1988) and Torgersen and others (1990) have presented a simple model for aqueous ^{222}Rn concentrations assuming secular equilibrium in the rock for ^{238}U to ^{226}Ra and describing radon emanation in terms of an emanating efficiency, E, U-content of rock, A_U, recoil distance, R_s, and specific surface area, ŝ. Their equation for radon, A_{Rn}, in a porefilling medium is:

$$A_{Rn} = A_U \times R_s \times E \times ŝ \quad (1)$$

The parameter ŝ would be difficult to measure exactly and may vary over several orders of magnitude, so reliable estimation would be similarly difficult. Thus, equation (1) may be applicable only in well-characterized systems (Torgersen, 1990).

Andrews and others (1982) showed a relationship between uranium content of the rock and radon content of the water that was derived for ground water in the Stripa granite of Sweden. Their equation (4) is similar to the one derived below, with some major differences. Most important, these authors assumed that all radon produced in the rock is transferred to the water. This assumption probably leads to large errors, as probably only a small portion of the radon is transferred to the water (Flügge and Zimens, 1939; this study). Further, several factors appear in their equation the origin of which is uncertain because no derivation is shown. It appears that these factors refer to specific properties of the Stripa granite, thus preventing general applicability of the model of Andrews and others (1982).

The models described above have the common trait that they are derived on the basis of microscopic properties of the system and therefore require measurements (or estimates) of these properties. Because of this dependence on microscopic properties, considerable uncertainty is introduced when applying these models to new systems that may not be very well characterized. Therefore it is desirable to have a generalized macroscopic model describing the potential for high radon concentrations in ground water.

GENERATION OF RADON IN ROCKS AND TRANSFER TO GROUND WATER

A model is derived for the simplified case in which radionuclides in the ^{238}U series are in secular equilibrium in the water-rock system within the sphere of influence of a pumping well. Then:

$$\lambda_U C_U = \lambda_{Rn} C_{Rn}, \quad (2)$$

where λ represents the radioactive decay constant (time^{-1}), C represents the number of atoms present in the water-rock system, and the subscripts U and Rn denote the radionuclides ^{238}U and ^{222}Rn. Note that in equation (2) no requirement is placed on any geometric property of the system, and no particular spatial distribution of uranium or radium in the rock is required. In fact, as shown later in this chapter, the results of this model may help derive some conclusions about the physical location of these elements in the rock. Equation (2) can be rearranged as follows:

$$\left(\frac{\lambda_U}{\lambda_{Rn}}\right) \times C_U = (C_{Rn})_{rock} + (C_{Rn})_{water}, \quad (3)$$

where all symbols are as defined before, but radon in rock and water are now accounted for separately. In a unit volume V_t of rock plus water with a porosity ϕ, the volume of rock equals $V_t(1-\phi)$ and the volume of water equals $V_t\phi$. Equation (3) can be recast, assuming that $[^{238}U]_w \ll [^{238}U]_r$. This assumption is justified noting that ^{238}U concentrations seldom exceed 10 ppb by mass in natural waters, whereas ^{238}U in most rocks is usually on the order of parts per million by mass. For a rock-water system with a porosity of 5%, the number of moles of uranium in the rock is more than 10^3 times that in the water. Thus neglecting dissolved uranium, noting that uranium concentrations in rock are commonly reported as parts per million by mass, and converting to mole quantities of each element leads to the following expression:

$$\left(\frac{\lambda_U}{\lambda_{Rn}}\right) \times \rho \times V_t(1-\varnothing) \times 4.2 \times 10^{-6} \times [U]_{rock} = \frac{(C_{Rn})_{rock} + (C_{Rn})_{water}}{A}, \quad (4)$$

where ρ represents the bulk density in kilograms per cubic decimeter, the factor 4.2×10^{-6} converts milligrams to moles of uranium, and A is Avogadro's number, 6.023×10^{23} mole^{-1}.

A factor can be introduced to describe the emanating efficiency of the rock with respect to release of radon to the ground water (Flügge and Zimens, 1939). This factor, E, is defined as:

$$E = \frac{(^{222}Rn)_{water}}{(^{222}Rn)_{water} + (^{222}Rn)_{rock}}. \quad (5)$$

Introduction of the factor E to the right-hand side of equation (4) results in:

$$\left(\frac{\lambda_U}{\lambda_{Rn}}\right) \times \rho \ V_t \ (1 - \emptyset) \times 4.2 \times 10^{-6} \times [U]_{rock} = \frac{(C_{Rn})_{water}}{E \times A}. \quad (6)$$

Noting that radon concentrations in water commonly are expressed in terms of picocuries per liter rather than moles, appropriate conversion factors can be introduced. Doing so leads to the following expression after rearrangement:

$$337 \times \rho \ V_t \left(\frac{1-\emptyset}{\emptyset}\right) \times E \times [U]_{rock} = (Rn)_{water}, \quad (7)$$

where uranium is expressed as parts per million by mass of rock, radon is expressed as picocuries per liter of water, E and ϕ are dimensionless, V_t is in cubic decimeters, and ρ is in kilograms per cubic decimeter.

APPLICATION OF THE MODEL AND COMPARISON TO FIELD DATA

Equation (7) can be used to predict radon activities in ground water based on the uranium concentration of the aquifer rock. Using this equation, a family of curves is calculated depending on the rock density, porosity, and value for E, as shown in Figure 1. As seen in the figure, for a given E value, porosity has the strongest effect on the result of the calcu;lation, with a minor effect of rock density. For a given U concentration in a rock (parts per million by mass), higher radon concentrations in ground water are expected in rocks of lower porosity, higher density, or higher emanating efficiency.

When comparing the model-generated curves to actual field data, as in Figure 2, it becomes apparent that when given an E value of 1, the model predicts ground-water ^{222}Rn values that are too high by a factor of 10 or more. Examination of equation (7) should yield some clues as to which conditions and assumptions should be questioned. The assumption that the water-rock system as a whole is in secular equilibrium is reasonable as shown by the data of Andrews and Woods (1972), Thamer and others (1981), Gundersen and others (1988), and Figure 3, which show that in the rock, ^{238}U and ^{226}Ra remain very close to secular equilibrium. Thus, only a few weeks' contact of ground water with the rock would be sufficient to attain secular equilibrium between ^{226}Ra and ^{222}Rn. Andrews and others (1990) pointed out that the residence time of ground water in a rock usually exceeds several half-lives of ^{222}Rn. The assumption of secular equilibrium between ^{238}U and ^{222}Rn therefore seems justified.

A more questionable condition is that all the ^{222}Rn generated in the system enters the ground water, i.e., that the value of E is 1. A significant proportion of the uranium and radium in the

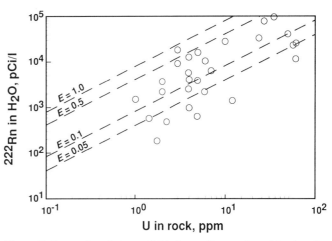

Figure 2. Comparison between field observations and model values for dissolved ^{222}Rn as a function of whole-rock U concentration. The model values were calculated for a porosity of 0.1 and average density of 2.7, using the E values adjacent to each line. Data are from crystalline aquifers in Pennsylvania, New Jersey, and Maryland (Wanty and Gundersen, 1988; Wanty and others, 1991a) and Colorado (Lawrence, 1990).

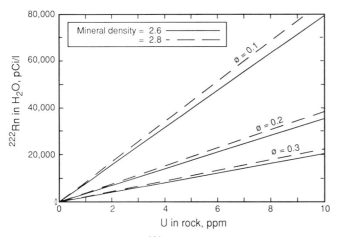

Figure 1. Calculated values of ^{222}Rn in water as a function of uranium concentration of the rock, using equation (7) and assuming $E = 1$ (see equation 5). Solid lines are for $\rho = 2.6$ g/cm^3, dashed lines are for $\rho = 2.8$ g/cm^3, ϕ = porosity.

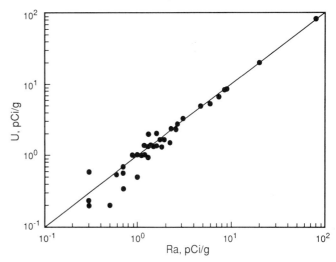

Figure 3. Data for uranium and radium in rock samples from crystalline rocks of the Appalachian Province and the Rocky Mountains. The line shows the 1:1 correlation that would be expected if the two radionuclides were in secular equilibrium.

rock must be located near the water-rock interface, or the average grain size of the rock must be less than the recoil distance of ^{222}Rn to produce ^{222}Rn activities as high as are observed relative to ^{226}Ra and ^{238}U (Wathen, 1987). Nevertheless, the reason for the deviation between predicted and observed values must be due to retention and attenuation of ^{222}Rn in the rock. Therefore, E is most likely much less than 1 in most systems. Recent work by Krishnaswami and Seidemann (1988) has shown that any nanopore network that exists in mineral grains would not lead to efficient transport of ^{222}Rn to ground water. Therefore, alpha-recoil and diffusion must be the key processes in transport of ^{222}Rn from the rock to the ground water. The fact that diffusion of ^{222}Rn through minerals is slow (Tanner, 1980) dictates that virtually all of the ^{222}Rn in the ground water must enter the water via direct alpha recoil (cf. Flügge and Zimens, 1939; Fleischer, 1988; Semkow, 1990; Semkow and Parekh, 1990).

The emanating efficiency, E. Although the factor E is simply defined (Equation [5]), it carries many implications for the location of uranium and radium in the rock (cf. Semkow, 1990). If most of the uranium and radium are dispersed throughout the lattices of accessory minerals, lower E values would be attained. Conversely, if these elements are localized at the water-rock interface, a higher value of E is expected. However, even if all of the U and Ra is at the water-rock interface, some Rn that is produced would be injected deeper into the aquifer rock by alpha recoil. Flügge and Zimens (1939) suggested a maximum upper limit of 0.5 for E for the limiting case in which all emanating atoms are located at the water-rock interface, noting that half of the vectors radiating from a point on the surface of a solid must be directed back into the solid. Thamer and others (1981) found emanating efficiencies as high as 0.55, but most of their values were less than 0.2 in dry systems and less than 0.3 in wet systems. Davis and others (1987) found very high degrees of radon loss (>80%) for some rock samples, but found lower values in coarser samples. Their "radon loss" parameter was based solely on gamma spectroscopy measurements of radon parents and daughters and would have benefited from chemical U analyses to compare to their effective U (eU) measurements.

Data of this study (Fig. 2) show that, unless unreasonably high porosities are assumed, values of E are almost always less than 0.5, and usually less than about 0.3. It should be borne in mind the curves in Figure 2 were calculated assuming a porosity of 10%, which is at the high end of the range expected for crystalline rocks (Freeze and Cherry, 1979). If more reasonable values of porosity in the range of 0 to 5% were used, lower E values would be required to fit the field data. The implication of this result is that a significant fraction of the emanating atoms must be located at a depth within the solid that is greater than the average alpha-recoil distance of ^{222}Rn (cf. Flügge and Zimens, 1939).

Limitations of the model. The model is derived for the general case, taking advantage of the averaging effects of a macroscopic view of the system. The most limiting assumption of the model is the requirement of secular equilibrium within the entire water-rock system. For systems not at secular equilibrium, a factor expressing that deviation needs to be incorporated. Calculations indicate that secular equilibrium between ^{238}U and ^{226}Ra is attained to within 5% in less than 2×10^6 yr if only ^{238}U is present initially and all daughters grow in from initial amounts of zero. Thus, the system must have exhibited closed behavior with respect to the radionuclides between ^{238}U and ^{226}Ra for that period of time.

A major advantage of the model derived above is that it is described in terms of parameters that are commonly available or measurable, such as density, porosity, and uranium concentration of the rock. However, more fundamental descriptions of the emanation efficiency may arise from models that consider microscopic aspects of the geometry of the water-rock interface (cf. Semkow, 1990; Semkow and Parekh, 1990) or from those described in terms of surface area of rock per volume of pore space (cf. Key and others, 1979; Bossus, 1984; Hammond and others, 1988; Torgersen and others, 1990), but such parameters are difficult or impossible to measure reliably (cf. Wanty and others, 1991b).

Another limitation of this model is the dependence of the results on the value of E. As discussed above, the value of E carries many implications for physical and chemical properties of the system. Anyone who uses this model must be aware of those implications as well as the problems and opportunities that they present. Many of the factors embodied in the E parameter are readily apparent when comparing this model with the microscopic models mentioned elsewhere in this chapter.

One case in which this model as presented breaks down is when a significant portion of the dissolved ^{222}Rn is derived from dissolved ^{226}Ra. This case represents a possible deviation from secular equilibrium between ^{238}U and ^{226}Ra in the rock. This situation is rare in natural, nonthermal, dilute waters such as most shallow ground-water wells. If such a case is encountered, two alternate solutions are possible. One would be to introduce a correction factor for the deviation from secular equilibrium. This factor could be determined from isotopic measurements of uranium and radium in the water and rock. Another alternative would be to rederive the equations, beginning after equation (3), to include a term for dissolved uranium.

Although the derivation of equation (7) and the ensuing discussion are couched in terms of radon emanation from rock to ground water, there is no mathematical requirement in the model for ground water to be filling pores. Thus, the model is applicable to the unsaturated zone as well. In this case, no uranium is expected to be in the soil gas, so the condition of $U_{rock} \gg U_{water}$ would be true. Conversely, different emanating efficiencies would be expected for dry rocks, as discussed above.

PREDICTIVE CAPABILITY GAINED FROM THE MODEL

Equation (7) can be used to calculate the radioactivity of radon in ground water in a rock of known uranium concentration, porosity, density, and emanating efficiency. A predictive capability may be gained, noting that reasonable limits may be placed on these variables in natural systems. For crystalline rocks, porosities less than 0.1 are expected (Freeze and Cherry, 1979). As seen in Figure 1, density of the rock does not significantly affect the result of the calculation for the range of values commonly observed. Thus, if the U concentration of the rock is known, a practical upper limit for the radon concentration can be calculated using equation (7) with a value of 0.5 for E. More likely, lower E values can be used to constrain the calculation further. Use of a lower E value may be justified by determining a representative value from the same aquifer by measuring U in the rock and ^{222}Rn in the ground water.

Equation (7) also can be applied to the problem of calculating the minimum U concentration necessary in a rock to produce an observed ^{222}Rn value. This technique may be applied to exploration for U deposits, as shown by Andrews and Wood (1972). As before, reasonable values of porosity and density can be assumed. A minimum value of U in the rock can be estimated using a value of 0.5 for E; more realistic values of U can be estimated using lower E values. As an example, a ground-water sample collected near Lyons, Colorado, was reported to have a ^{222}Rn value of 2.5×10^6 pCi/L (P. Nyberg, personal communication, 1989). Using $\phi = 0.1$, $E = 0.5$, and $\rho_r = 2.6$ g/cm^3, a minimum U concentration of 314 ppm by mass is calculated for the rock. Using the more reasonable value of $E = 0.1$, the U concentration must be five times greater.

An interesting application of the model derived here may be to aid estimates of specific surface area of the rock. If the porosity, density, and uranium content of the rock are measured, as well as the ^{222}Rn content of the water with which the rock is in contact, then the emanating efficiency can be calculated. The value for E could then be used in equation (1) (Hammond and others, 1988) or in an equation such as that presented by Bossus (1984) to calculate the specific surface area of the rock, $\hat{s}$. Torgersen and others (1990) have equated this value of $\hat{s}$ with that of Pačes (1973), although this is not likely to be the case, as discussed by Wanty and others (1991). The value of $\hat{s}$ calculated by this approach represents the total surface area, whereas the value of $\hat{s}$ calculated by Pačes' (1973) model is the reactive surface area of feldspar grains, a subset of the former.

CONCLUSIONS

A model has been developed to calculate the expected value of radon in ground water, if the U concentration of the rock, the porosity, the density, and the emanating efficiency of the rock are known. Although values for these parameters are not always known, reasonable values can be assumed to constrain the result of the calculation. Alternatively, well-characterized samples could be used to calibrate the model for each rock type in a study area. From a comparison of the model to data from crystalline rocks, it is apparent that E is seldom greater than 0.5, and usually is less than 0.3. Thus, most of the radon produced in a water-rock system never reaches the water before it decays. The model can be applied to the prediction of values of radon in a ground water if the U concentration of the rock is known, or conversely to the problem of estimating the U concentration of the rock if the radon value is known for the ground water. Further work remains to be done to characterize E values for sedimentary rock systems, and to more tightly constrain E values for crystalline rocks.

ACKNOWLEDGMENTS

Funding for this work was provided by the U.S. Geological Survey and by the U.S. Department of Energy. We thank Francis Hall and an anonymous reviewer for their criticisms of this manuscript.

REFERENCES CITED

Andrews, J. N., and Wood, D. F., 1972, Mechanism of radon release in rock matrices and entry into groundwaters: Transactions of the Institute of Mining and Metallurgy, Section B, B198–B209.

Andrews, J. N., and 8 others, 1982, Radioelements, radiogenic helium, and age relationships for groundwaters from the granites at Stripa, Sweden: Geochimica et Cosmochimica Acta, v. 46, p. 1533–1543.

Andrews, J. N., D. J. Ford, N. Hussain, D. Trivedi, and M. J. Youngman, 1989, Natural radioelement solution by circulating groundwaters in the Stripa granite: Geochimica et Cosmochimica Acta, v. 53, p. 1791–1802.

Bossus, D.A.W., 1984, Emanating power and specific surface area: Radiation Protection Dosimetry, v. 7, p. 73–76.

Cothern, C. R., 1987, Estimating the health risks of radon in drinking water: Journal of the American Water Works Association, v. 79, p. 153–158.

Crawford-Brown, D.J., 1990, Analysis of the health risk from ingested radon, in Cothern, C. R., and Rebers, P. A., eds., Radon, radium, and uranium in drinking water: Chelsea, Michigan, Lewis Publishers, p. 17–26.

Cross, F. T., Harley, N. H., and Hofmann, W., 1985, Health effects and risks from ^{222}Rn in drinking water: Health Physics, v. 48, p. 649–670.

Davis, N. M., R. Hon, and Dillon, P., 1987, Determination of bulk radon emanation rates by high resolution gamma-ray spectroscopy, in Graves, B., ed., Radon in ground water: Chelsea, Michigan, Lewis Publishers, p. 111–129.

Fleischer, R. L., 1983, Theory of alpha recoil effects on radon release and isotopic disequilibrium: Geochimica et Cosmochimica Acta, v. 47, p. 779–784.

——, 1988, Alpha-recoil damage: Relation to isotopic disequilibrium and leaching of radionuclides: Geochimica et Cosmochimica Acta, v. 52, p. 1459–1466.

Flügge, S., and Zimens, K. E., 1939, Die Bestimmung von Korngrössen und Diffusionskonstanten aus dem Emaniervermögen (Die Theorie der Emaniermethode): Zeitschrift für Physikalische Chemie (Leipzig), B, 42, p. 179–220.

Freeze, R. A., and Cherry, J. A., 1979, Groundwater: Englewood Cliffs, New Jersey, Prentice-Hall, 604 p.

Gundersen, L.C.S., Reimer, G. M., and Agard, S. S., 1988, Correlation between geology, radon in soil gas, and indoor radon in the Reading Prong, in Marikos, M. A., and Hansman, R. H., eds., Geological causes of natural radionuclide anomalies, Missouri Department of Natural Resources Special Publication 4, p. 91–102.

Hammond, D. E., Leslie, B. W., Ku, T.-L., and Torgersen, T., 1988, ^{222}Rn concentrations in deep formation waters and the geohydrology of the Cajon

Pass borehole: Geophysical Research Letters, v. 15, p. 1045–1048.

Hess, C. T., Vietti, M. A., Lachapelle, E. B., and Guillemette, J. F., 1990, Radon transferred from drinking water into house air, *in* Cothern, C. R., and Rebers, R. A., eds., Radon, radium, and uranium in drinking water: Chelsea, Michigan, Lewis Publishers, p. 51–68.

Key, R. M., Guinasso, N. L., Jr., and Schink, D. R., 1979, Emanation of radon-222 from marine sediments: Marine Chemistry, v. 7, p. 221–250.

Krishnaswami, S., and Seidemann, D. E., 1988, Comparative study of ^{222}Rn, ^{40}Ar, ^{39}Ar, and ^{37}Ar leakage from rocks and minerals: Implications for the role of nanopores in gas transport through natural silicates: Geochimica et Cosmochimica Acta, v. 52, p. 655–658.

Lawrence, E. P., 1990, Hydrogeologic and geochemical processes affecting the distribution of ^{222}Rn and its parent radionuclides in ground water, Conifer, Colorado: Golden, Colorado School of Mines (unpublished M.S. thesis), 181 p.

Lawrence, E. P., Wanty, R. B., and Nyberg, P. A., 1992, Contribution of ^{222}Rn in domestic water supplies to ^{222}Rn in indoor air in homes in Colorado: Health Physics, v. 62, p. 171–177.

Nazaroff, W. W., and Nero, A. V., Jr., 1988, Radon and its decay products in indoor air, 518 p., New York, John Wiley & Sons, 518 p.

Nazaroff, W. W., Doyle, S. M., Nero, A. V., Jr., and Sextro, R. G., 1988, Radon entry via potable water, *in* Nazaroff, W. W., and Nero, A. V., Jr., eds., Radon and its decay products in indoor air: John Wiley & Sons, p. 131–160.

Paĉes, T., 1973, Steady-state kinetics and equilibrium between ground water and granitic rock: Geochimica et Cosmochimica Acta, v. 37, p. 2641–2663.

Prichard, H. M., and Gesell, T. F., 1981, An estimate of population exposures due to radon in public water supplies in the area of Houston, Texas: Health Physics, p. 599–606.

Semkow, T. M., 1990, Recoil-emanation theory applied to radon release from mineral grains: Geochimica et Cosmochimica Acta, v. 54, p. 425–440.

Semkow, T. M., and Parekh, P. P., 1990, The role of radium distribution and porosity in radon emanation from solids: Geophysical Research Letters, v. 17, p. 837–840.

Tanner, A. B., 1980, Radon migration in the ground: A supplementary review, *in* Gesell, T. F., and Lowder, W. M., eds., Natural radiation environment III, Symposium Proceedings, U.S. Department of Energy Report CONF-780422, p. 5–56.

Thamer, B. J., Nielson, K. K., and Felthauser, K., 1981, The effects of moisture on radon emanation: U.S. Bureau of Mines Open-File Report 184–82.

Torgersen, T., Benoit, J., and Mackie, D., 1990, Controls on groundwater Rn-222 concentrations in fractured rock: Geophysical Research Letters, v. 17, p. 845–848.

Wanty, R. B., and Gundersen, L.C.S., 1988, Groundwater geochemistry and radon-222 distribution in two sites on the Reading Prong, eastern Pennsylvania, *in* Marikos, M. A., and Hansman, R. H., eds., Geological causes of natural radionuclide anomalies: Missouri Department of Natural Resources Special Publication 4, p. 147–156.

Wanty, R. B., Johnson, S. L., and Briggs, P. H., 1991a, Radon-222 and its parent radionuclides in ground water from two study areas in New Jersey and Maryland: Applied Geochemistry, v. 6, p. 305–318.

Wanty, R. B., Rice, C. A., Langmuir, D., Briggs, P. H., and Lawrence, E. P., 1991b, Prediction of uranium adsorption by crystalline rocks: The key role of reactive surface area: Materials Research Society Symposium Proceedings Series, v. 212, p. 695–702.

Wathen, J. B., 1987, The effect of uranium siting in two-mica granites on uranium concentrations and radon activity in ground water, *in* Graves, B., ed., Radon in ground water: Chelsea, Michigan, Lewis Publishers, p. 31–46.

MANUSCRIPT ACCEPTED BY THE SOCIETY APRIL 6, 1992

The influence of season, bedrock, overburden, and house construction on airborne levels of radon in Maine homes

E. Melanie Lanctot
Division of Disease Control, Maine Department of Human Services, 157 Capitol Street, State House Station 11, Augusta, Maine 04333

Peter W. Rand and Eleanor H. Lacombe
Research Department, Maine Medical Center, 22 Bramhall Street, Portland, Maine 04102

C. Thomas Hess
Department of Physics, University of Maine, Orono, Maine 04469

Gregory F. Bogdan
Division of Disease Control, Maine Department of Human Services, 157 Capitol Street, State House Station 11, Augusta, Maine 04333

ABSTRACT

As part of a case-control study conducted to test the association between indoor air radon levels and lung and other cancers, 539 air and 547 water radon samples were measured from homes with drilled wells throughout Maine. The factors tested for an association with air radon levels were season, house construction, draftiness of house, heating source, basement construction, radon levels in the water supply, and the type of overburden and bedrock underlying the house. Logistic regression was used to determine which variables were significantly associated with indoor air radon levels of 4 pCi/L or above and to assess their relative importance. House construction, water radon levels, and overburden permeability showed the strongest association with air radon levels. Woodframe homes were 10 times more likely than mobile homes to have an air radon level of 4 pCi/L or more (95 percent confidence interval [CI] = 1.1 to 103), and brick or stone houses were 108 times more likely to have high air radon levels (CI = 6.7 to 1,760). Homes with water radon between 10,001 and 100,000 pCi/L were 6 times more likely to have high air radon (CI = 1.6 to 22.7), and homes with water radon levels above 100,000 pCi/L were 54 times more likely to have air levels of 4 pCi/L or above than were homes with water radon at or below 1,000 pCi/L (CI = 1.9 to 1,529). Homes built over sand and gravel were 12 times more likely to have high air radon than homes built over clay (CI = 4.1 to 37.1), and homes built over two-mica granite were 4 times more likely to have high air radon than homes built over low-grade metamorphic rock (CI = 1.4 to 11.5). In addition, high air radon levels were more likely in the winter than in the summer (odds ratio [OR] = 2.9, CI = 1.3 to 6.2) and in homes heated with electricity than with other sources of heat (OR = 3.5, CI = 1.4 to 8.8).

INTRODUCTION

Radon was first detected in Maine in 1957 in ground water extracted from a granite pluton (Grune and others, 1960). As a result, for the next 20 yr studies in Maine focused on ground water as the principal source of airborne radon in homes (Hoxie, 1966; Hess and others, 1979). However, the focus nationally has been on other sources of indoor radon. High radon concentrations (i.e., activity due to radon decay) have been observed in homes built over phosphate tailings in Florida (Roessler and

---

Lanctot, E. M., Rand, P. W., Lacombe, E. H., Hess, C. T., and Bogdan, G. F., 1992, The influence of season, bedrock, overburden, and house construction on airborne levels of radon in Maine homes, *in* Gates, A. E., and Gundersen, L.C.S., eds., Geologic Controls on Radon: Boulder, Colorado, Geological Society of America Special Paper 271.

others, 1983) and over mylonitized gneiss near Boyertown, Pennsylvania (Reimer and Gundersen, 1989). In Sweden, alum shale incorporated into aerated concrete used for building materials was found to contribute significant amounts of radon to indoor air (Åkerblom and Wilson, 1981). The air radon screening protocol of the U.S. Environmental Protection Agency (EPA) reflects the concern that gas from the soil and bedrock, rather than from the water supply, is the primary source of indoor radon by the recommendation that air radon be tested in the lowest livable level of the home, which may not be the location where the water use is the heaviest (U.S. Environmental Protection Agency, 1986).

The problem of predicting which homes will have high air radon concentrations is complicated by the fact that there are a number of secondary factors that influence the amount of radon entering the home and the degree to which it can accumulate. These include the construction of the house (George and Breslin, 1980; Radford, 1985), the construction of the basement (Mushrush and Mose, 1988; Centers for Disease Control, 1989; Buchli and Burkart, 1989), the draftiness of the house (Hess and others, 1985; Cohen and Gromicko, 1988), and the season of the year (Wilkening and Wicke, 1986; Mushrush and Mose, 1988; Hess and others, 1985).

In none of these studies were the data analyzed to estimate quantitatively the incremental radon risk associated with each characteristic measured. This chapter presents an analysis of the individual effects of differences in bedrock and surficial geology, season, and characteristics of the house on the concentration of indoor air radon, and a unique application of logistic regression to measure the relative influence of these factors on the odds of having a house with a radon level at or exceeding the EPA's screening level of 4 pCi/L.

METHODS

Data collection. Data were collected as part of a cooperative case control study to determine the association between indoor air radon levels and lung and other cancers. Because of the concern about high levels of radon in the water supply, participation was limited to individuals who had lived in Maine for at least 10 yr in a home served by a private well drilled into bedrock. Each participant completed a questionnaire that, in addition to demographic, health, and exposure information, asked for specific details concerning the construction, water source, heating method, and draftiness of each home. The location of dwellings was plotted on bedrock and surficial geologic maps at the largest scale available (Osberg and others, 1985; Thompson and Borns, 1985).

One member of the study team visited all homes in the study to take water radon measurements and leave air monitoring devices. For water radon sampling, tap water from the kitchen faucet was flushed for at least 5 min, and duplicate 10-ml aliquots of bubble-free water were transferred by syringe into two 25-ml scintillation vials containing 5 ml of mineral oil fluor. The vials were transported directly to the Radiation Physics Laboratory at the University of Maine for radon analysis by a modification of the method developed by Prichard and Gesell (1978). To measure the air radon, one alpha-track detector was left for 2 to 3 mo on the top of the refrigerator. This location was selected because it standardized, to the extent possible, the location of the cup, and was near a significant water source, the kitchen sink. Type F cups used in the initial 2 yr of the project were eventually replaced by the smaller and more robust SF monitors. In 45 homes F and SF cups were exposed simultaneously to determine if there were systematic differences between the two types of measurements.

Data analysis. The distribution of the air and water radon values are highly skewed. For this reason, median radon levels were calculated and nonparametric analyses of variance were used to test whether measured radon values for two or more populations were significantly different. Specifically, the Wilcoxon (1945) test was used when two classifications of a variable were compared (e.g., summer vs. winter), and the Kruskal-Wallis (1952) test was used when more than two classifications were compared (e.g., sand, till, and clay overburden). When the measurements were paired (winter vs. summer in the same home; winter values for successive years in the same home), the geometric means were calculated, and paired t-tests were used to determine the statistical significance of the difference between the means. The probability (p) shown in Tables 1 through 3 indicates the level of confidence that there is a true difference in the radon values among the classifications compared. The generally accepted level of statistical significance is $p < 0.05$, which means that, if there were no actual difference among these classifications, the chance of seeing a difference as large as that observed is less than 5 in 100.

Logistic regression, traditionally used to estimate the relative strength of the effect of risk factors on the odds of contracting a disease, is used in this study to determine the relative importance of risk factors on the odds of having a house with air radon concentrations at or above EPA's standard of 4 pCi/L (see Breslow and Day, 1980, for a discussion of this statistical method). The formula for the multivariate model of the logistic regression is:

$$\log (P/1 - P) = \alpha + \beta_1 X_1 + \beta_2 X_2 + \ldots \beta_n X_n,$$

where P = probability of a positive outcome (i.e., having air radon at or exceeding 4 pCi/L^{-1}), α = log odds for homes with zero for all predictors included in analysis, β_n = log odds for the effect of variable X_n, and X_n = risk variable.

In the logistic regression, each classification of a risk factor with higher median air radon values was compared to the classification with the lowest median air radon value. The risk factors used in this analysis were season (winter vs. summer), house construction (woodframe and brick/stone houses vs. mobile homes), basement (present vs. absent), basement floor construction (concrete vs. other), draftiness (tight vs. drafty), heat source (electric vs. other), overburden type (till and sand/gravel vs.

TABLE 1. MEDIAN WATER AND AIR RADON LEVELS BY TYPE OF BEDROCK

	Water Radon			Air Radon		
	Number	Median (pCi/L)	Range (pCi/L)	Number	Median (pCi/L)	Range (pCi/L)
Metamorphic rock						
Low grade	138	1,579	85–18,115	137	1.40	0.10–9.05
Medium grade	94	2,561	264–46,103	94	1.39	0.20–9.58
High grade	153	2,969	254–65,656	151	1.54	0.14–21.50
Plutonic rock						
Two-mica granite	99	9,918	142–151,182	98	3.10	0.27–24.75
Other	57	3,326	341–72,229	54	1.00	0.20–7.14
Unknown*	6	4,990	264–8,1115		1.87	0.70–5.49
		$p = 0.0001$			$p = 0.0001$	

*Location of home on map too vague to determine rock type.

TABLE 2. INDOOR AIR RADON BY OVERBURDEN TYPE AND SEASON

	Summer			Winter		
	Number	Median (pCi/L)	Range (pCi/L)	Number	Median (pCi/L)	Range (pCi/L)
Sand/gravel	24	1.43	0.27–7.14	31	1.94	0.28–23.44
Till	88	1.45	0.14–13.28	229	1.70	0.10–24.75
Clay	27	1.42	0.27–7.71	95	1.15	0.19–9.05
		$p = 0.64$			$p = 0.0001$	

clay), bedrock type (medium-grade metamorphic rock, high-grade metamorphic rock, 2-mica granite, and other plutonic rock vs. low-grade metamorphic rock), and water radon concentration (1,001 to 10,000 pCi/L, 10,001 to 100,000 pCi/L, and >100,000 pCi/L vs. ≤1,000 pCi/L). In the first analysis, each risk factor was fitted to a univariate model. Those factors with a statistically significant or nearly significant odds ratio were then fitted to a multivariate model, in which the log odds for each factor is calculated with the other factors held constant.

RESULTS

F vs. SF track etch detectors. For the 45 homes where air radon levels were measured with both type F and type SF devices, the geometric means of the two devices were not significantly different: $GM_F = 2.72$ pCi/L, $GM_{SF} = 3.07$ pCi/L, $p = 0.19$. As a result, we believe that the change in cup type has not biased the results discussed below.

Radon distribution. Five hundred thirty-nine air radon measurements and 547 water measurements were collected for this study. The linear and log-normal distributions and arithmetic and geometric means and medians are given in Figure 1.

In 16.1 percent of the homes, water radon concentrations exceeded 10,000 pCi/L, 8.6 percent exceeded 20,000 pCi/L, and 0.7 percent were over 100,000 pCi/L. The distribution of air radon levels showed 13.4 percent exceeding 4 pCi/L.

Studies have shown that, in homes with no other radon source and one air exchange per hour, the exchange coefficient for radon between water and air is approximately 10,000:1 (Hess and others, 1982; Kahlos and Asikainen, 1980). This relationship is shown by the dashed line in Figure 2, which plots individual winter and summer air-water radon pairs (n = 235, points densely clustered near origin were omitted from the figure). Points above the line reflect homes with either low ventilation rates or non-water radon sources (such as the underlying soils), whereas points below the line indicate homes with increased ventilation (as in the summer) or low water use. The clustering of data points along the ordinate suggest that much of the air radon did not come from the water supply but rather from another source, most likely the underlying soils. The proportion of tight to drafty homes (approx-

TABLE 3. AIR RADON BY HOUSE CHARACTERISTICS

	Number	Median (pCi/L)	Range (pCi/L)
House Construction			
Woodframe	490	1.60	0.10–24.75
Mobile	36	1.00	0.35–7.00
Brick/stone	11	4.07	0.31–23.44
Unknown	2	1.11	0.42–1.80
		$p = 0.0002$	
Foundation			
Concrete/cinder block	288	1.66	0.10–24.75
Rock, unspecified	105	1.55	0.31–23.44
Granite	69	1.58	0.20–12.13
Other/unknown/none	77	1.16	0.21–17.34
		$p = 0.16$	
Basement Floor			
Concrete	304	1.70	0.10–24.75
Dirt	136	1.57	0.20–12.13
Stone/granite	32	1.12	0.20–5.69
Unknown	67	1.01	0.21–17.34
		$p = 0.03$	
Heat Source			
Electric	43	1.70	0.20–14.70
Other	492	1.54	0.10–24.75
Unknown	4	1.34	0.42–3.00
		$p = 0.07$	
Draftiness			
Drafty	215	1.58	0.20–12.13
Tight	320	1.57	0.10–24.75
Unspecified	4	3.31	0.42–4.99
		$p = 0.5$	

was more variation when radon levels were high, radon levels that were elevated in the initial measurement were generally high when remeasured. Contributing to the variability at low radon levels are errors inherent in the imprecise alpha-track method, which can exceed 20 percent for radon levels of less than 1 pCi/L and 1-yr exposure (R. A. Ostwald, personal communication, 1987).

Finally, the arithmetic mean, median, and upper percentiles of indoor air radon values for consecutive months were evaluated with the average monthly heating degree days for Portland, Maine (Fig. 5; the location of Portland is shown in Fig. 6). Summer values were used when both summer and winter measurements were taken in the same house. Seasonal differences are not as apparent with mean or median values as they are in the highest quartile of radon values; i.e., the homes with the highest radon values will show the greatest seasonal fluctuation. Although elevated values are found in the nonheating season, the highest values are found in the winter months and generally follow the heating degree day curve.

Geological influences

The data show a generally progressive increase in median water radon levels with increasing degree of metamorphic grade of the bedrock (Table 1). The highest median radon level occurred in ground water from two-mica granites, which was three times higher than from other plutonic rock (71% of which are other granites in our sample). Air radon was also significantly higher in homes built over two-mica granite. The average air radon for other plutonic rock was actually less than that for metamorphic rock. Figure 6 shows the generalized metamorphic grade of bedrock for the state of Maine and the location of the two-mica granites sampled for this study. A statistically significant association between overburden type and median indoor air radon was apparent only during the winter (Table 2). During the heating season, the median radon value in homes built over till was higher than in homes built over clay, and the radon in homes over sand and gravel was higher than in those over clay or till.

House characteristics

A summary of the relationship between house characteristics and air radon levels is given in Table 3. The median air radon in woodframe houses was 60 percent higher than in mobile homes. The median air radon in brick and stone houses was 250 percent greater than in woodframe homes and 400 percent greater than in mobile homes; however, the sample size of stone and brick houses and mobile homes was small. A statistically significant difference was also observed among different types of basement floor construction. The highest median levels were in homes that had basements with concrete floors, had concrete or cinderblock walls, or that used electricity as the primary source of heat. There was essentially no difference in median air radon levels between "tight" and "drafty" houses.

imately 3:2; see Table 3) argues against decreased ventilation as the primary cause of this clustering.

Seasonal effects. Of the 539 air radon measurements, 392 were taken during the heating season (median detector placement date between November 1 and April 15), and the remaining 147 were taken during the summer. The median winter and summer air radon concentrations were 1.59 and 1.48 pCi/L, respectively; there was no significant difference between the two groups of measurements (p = 0.37).

More specific information on seasonal influences in individual homes was provided by taking an additional summer measurement on 39 of the homes initially measured in the winter (Fig. 3). The geometric mean of the winter values was 12 percent greater than the summer values (1.27 vs. 1.13 pCi/L, respectively) but, again, the difference was not statistically significant (p = 0.42).

To assess the variability within one season, 25 homes with initial winter radon concentrations as high as 25 pCi/L were remeasured during the subsequent winter (Fig. 4). Although there

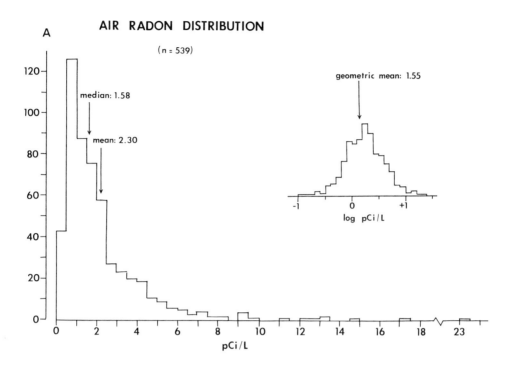
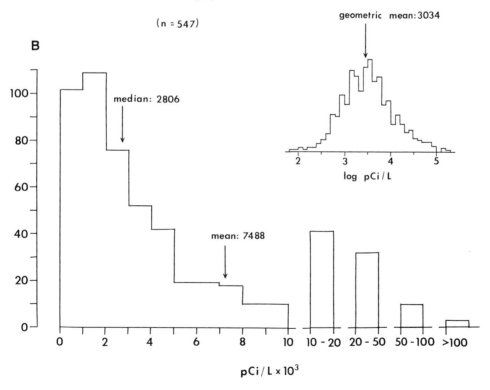

Figure 1. A, Distribution of radon values in air. B, Distribution of radon values in water.

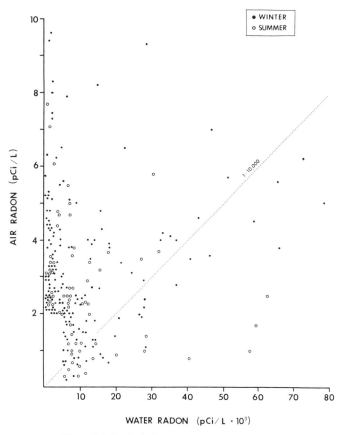

Figure 2. Matched air-water radon measurements.

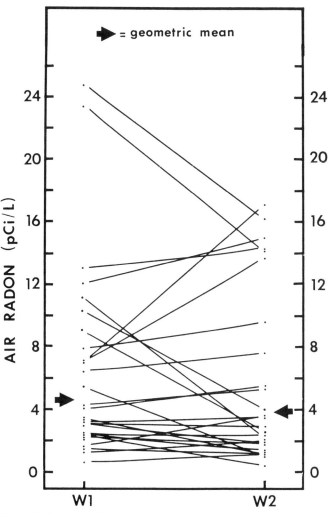

Figure 4. Repeated winter air radon values in 25 homes. Lines connect values in same homes.

Results of logistic regression analysis

The odds ratios (OR), probabilities (P), and 95% confidence intervals (CI) determined by logistic regression are shown in Table 4. When fitted to the univariate model, winter values were almost twice as likely to be 4 pCi/L or higher than were summer values, which approached statistical significance. However, when other significant variables were controlled, the odds ratio was almost 3 (CI = 1.3 to 6.2). Also, the odds of having high radon in woodframe homes tested against mobile homes was not significant until fitted to the multivariate model (OR = 10.4, CI = 1.1 to 103). Conversely, the draftiness of the house lost significance as a predictor of high air radon when fitted to the multivariate model. Homes with basements were twice as likely to have high air radon as homes without basements, although this difference was not statistically significant (CI = 0.8 to 5.2).

The highest risk was associated with homes built of brick or stone (OR = 108; CI = 6.7 to 1,760) although there were only 11

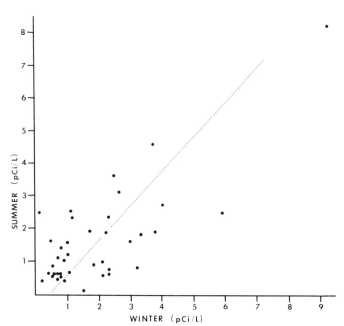

Figure 3. Summer vs. subsequent winter air radon levels in 39 homes.

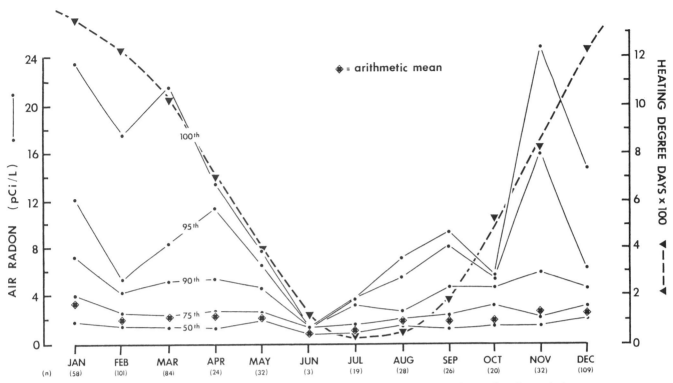

Figure 5. Mean and 50th, 75th, 90th, 95th, and 100th percentile points for median dates of air radon measurement, compared to average heating degree days by month for Portland, Maine (1870–1986).

such cases in this study. Electrically heated homes were 3.5 times more likely to have 4 pCi/L or more of air radon than were homes heated by other sources (CI = 1.4 to 8.8), and homes built over sand and gravel were 12 times more likely to have high radon levels than those built over clay (CI = 4.1 to 37.1). The only type of bedrock that significantly increased the risk of high air radon levels in the home was two-mica granite, whereas the risk of high radon in homes over other plutonic rock was not significantly different than in homes over low-grade metamorphic rock (CI = 0.3 to 4.3). Homes with water radon values between 10,001 and 100,000 pCi/L were six times more likely to have high air radon levels than were homes with a water radon level of 1,000 pCi/L or less (CI = 1.6 to 22.7), and the four homes with water radon above 100,000 pCi/L were 54 times more likely to have air radon levels of 4 pCi/L or above (CI = 1.9 to 1,529).

DISCUSSION

In other studies, average air radon measurements in the winter have been higher than summer measurements (Cohen and Gromicko, 1988; Hess and others, 1985; Mushrush and Mose, 1988; Wilkening and Wicke, 1986). In our study, season of the year did not show a significant association with air radon using a nonparametric test; however, the logistic regression, multivariate model, showed winter measurements to be nearly three times more likely to be 4 pCi/L or above than summer measurements. This is not contradictory because the nonparametric test compares the entire distribution of the air measurements, whereas the logistic regression compares the risk of having values at the upper tail of the distribution (13% of our air radon measurements exceeded 4 pCi/L). Figure 5 shows that there were higher measurements in the 95th and 100th percentiles in the winter months, whereas the median monthly values (50th percentile) varies little throughout the year.

The logistic regression also identified overburden type as one of the stronger predictors of which homes have high air radon, presumably because particle size is associated with permeability. However, the comparison of the distributions, stratified by season, showed that this effect is significant only in the winter.

The strong seasonal effect may be due to frost retarding the migration of radon through the soil to the atmosphere (Nazaroff and others, 1988), coupled with the pressure gradient created by the temperature differences between the inside and the outside of the house (the "stack effect"), which draws radon out of the soil (Buchli and Burkart, 1989; Wilkening and Wicke, 1986).

Although our data show that the median value of first-floor air radon measurements in homes with concrete floor basements were higher than in homes with dirt floor basements, the presence or absence of a basement was not a predictor of whether first floor concentrations would exceed EPA standard. One might expect that homes with dirt floor basements would actually have higher radon because there would be little to impede radon transfer from the soil. Buchli and Burkart (1989) did find that the

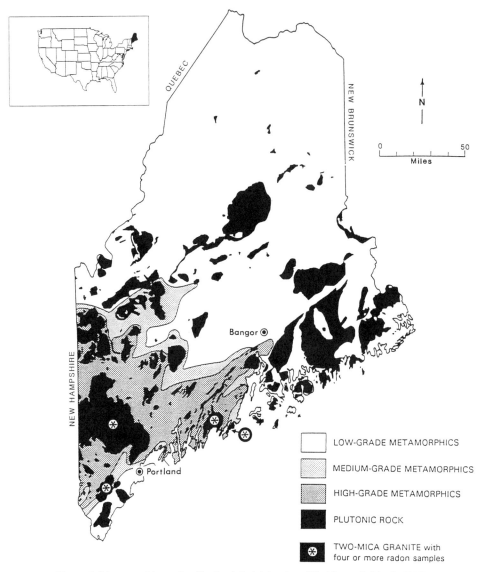

Figure 6. Metamorphic grade of bedrock in Maine (modified from Guidotti, 1985).

average radon level taken in cellars with partial gravel or earth floors was 5.4 times higher than in cellars with concrete floors. Our sample of homes with dirt floor basements may actually have higher basement radon concentrations, but less of the radon is reaching the upper levels of the house, probably because the basements are older and draftier.

Because our air radon measurements were taken on the first floor, our mean values may be two to three times lower than those found in other studies in which basement sampling is included in the protocol (George and Breslin, 1980; Hess and others, 1982; Fleischer and others, 1983).

Other studies have found that tighter houses have higher air radon (Buchli and Burkart, 1989; Hess and others, 1985); however, the tightness of the houses in our study, as subjectively assessed by the occupant, was not a significant predictor of high indoor radon. Perhaps a better indicator of tightness is the source of heat, which did show a significant association with high indoor radon (electrically heated homes in our study were 3.5 times more likely to have concentrations at or exceeding 4 pCi/L). Due to the cost of electricity, electrically heated homes are apt to be tighter, and fewer air exchanges result in the buildup of radon levels. And although oil and wood furnaces may draw more radon into basements, they may also be venting the radon up the chimney.

Data from our study indicate that homes built of brick or stone have high risk for elevated air radon concentrations relative to mobile homes, although there were only 11 brick/stone houses and 36 mobile homes sampled. It was not determined whether these building materials are the source of the radon or some other, untested, aspect of the construction of these houses is causing

TABLE 4. RELATIVE STRENGTH OF RISK FACTOR EFFECTS ON HAVING A HOUSE WITH AIR RADON ≥4 pCi/L: RESULTS OF LOGISTIC REGRESSION

Risk Factors	OR	P	95% CI
Univariate Model			
Season			
Winter vs. summer	1.9	0.055	1.0–3.5
House Construction			
Woodframe vs. mobile home	5.6	0.091	0.8–41.9
Brick/stone vs mobile home	42.0	0.002*	4.1–425
Basement			
With vs. without	2.0	0.150	0.8–5.2
Basement Floor			
Concrete vs. other	1.4	0.252	0.8–2.4
Draftiness			
Tight vs. drafty	1.8	0.034*	1.0–3.1
Heat			
Electric vs. other	2.6	0.008*	1.3–5.4
Overburden			
Till vs. clay	2.6	0.025*	1.1–5.9
Sand/gravel vs. clay	8.7	<0.001*	3.4–22.3
Bedrock			
Medium grade vs. low grade	2.1	0.113	0.8–5.2
High grade vs. low grade	1.5	0.400	0.6–3.5
Other plutonic vs. low grade	1.5	0.523	0.5–4.5
Two-mica granite vs. low grade	7.6	<0.001*	3.4–16.7
Water Radon			
>10^3 to 10^4 vs. ≤10^3 pCi/L	1.6	0.292	0.7–3.6
>10^4 to 10^5 vs. ≤10^3 pCi/L	5.6	<0.001*	2.3–13.7
>10^5 vs. ≤10^3 pCi/L	12.4	0.019*	1.5–102
Multivariate Model			
Season			
Winter vs. summer	2.9	0.007*	1.3–6.2
House Construction			
Woodframe vs. mobile home	10.4	0.045*	1.1–103
Brick/stone vs. mobile home	108.4	<0.001*	6.7–1760
Heat			
Electric vs. other	3.5	0.007*	1.4–8.8
Overburden			
Till vs. clay	1.8	0.245	0.7–4.5
Sand/gravel vs. clay	12.3	<0.001*	4.1–37.1
Draftiness			
Tight vs. drafty	1.2	0.544	0.6–2.3
Bedrock			
Medium grade vs. low grade	2.2	0.134	0.8–6.4
High grade vs. low grade	0.9	0.771	0.3–2.5
Other plutonic vs. low grade	1.1	0.859	0.3–4.3
Two-mica granite vs. low grade	3.9	0.012*	1.4–11.5
Water Radon			
>10^3 to 10^4 vs. ≤10^3 pCi/L	1.9	0.277	0.6–5.8
>10^4 to 10^5 vs. ≤10^3 pCi/L	6.0	0.009*	1.6–22.7
>10^5 vs ≤10^3 pCi/L	54.2	0.019*	1.9–1529

*Statistically significant, $p < 0.05$

more radon to enter or be retained in the house. Radford (1985, p. 283) stated that "stones and brick used for house walls can contain variable amounts of radium, and usually stone or brick houses have somewhat higher radon indoors than frame houses"; however, the median air radon concentration found by George and Breslin (1980) for first floor measurements in five brick homes in the New York City area was only 0.43 pCi/L (our median value was 4.07 pCi/L). Obviously, further measurements in brick and stone houses are warranted. The most likely explanation for the low air radon values seen in mobile homes is that these structures are often raised above the ground, resulting in a diminished contribution from radon in the soil.

As in our study, Hess and others (1983) found that water radon levels from bedrock wells correlated with increasing grade of bedrock metamorphism, and that the highest levels were in granitic rock. However, we found that two-mica granite contributes to the highest radon concentrations in both water and indoor air, and that other plutonic rock, primarily granites in our study, actually have lower median air radon levels than do low-grade metamorphic rock. Two-mica granites are commonly enriched in uranium as uraninite and coffinite, which are nearly all in a form readily transported by ground water (Hall and others, 1987). Ground water contributed significantly to air radon levels at a magnitude of 10^4 pCi/L or greater, which included about 16% of the homes sampled. Levels this high are most likely to occur in the two-mica granites. Therefore, in areas with abundant two-mica granites, a protocol for screening air radon must consider the level of the house where the water use is the heaviest. In some cases, depending on the permeability of the overburden, two screening tests may be appropriate, e.g., one in the basement and one in the kitchen.

There are several limitations to this study. The houses sampled were not randomly selected; nearly half the homes were occupied by cancer patients, and the controls were self-selected. Because participation in this study was limited to persons who had lived in a home with a privately owned well for at least 10 yr, homes built after the 1974 energy crisis were not eligible when we started collecting data in 1982. Therefore, houses in this sample may not be representative of newly built homes. There were no solar-heated homes represented by the data, and the sample size of mobile homes and of stone or brick houses was small, thus limiting the conclusions that can be drawn concerning the association between house construction and air radon levels.

CONCLUSIONS

1. In Maine, radon levels of 4 pCi/L or above in indoor air are three times more likely to occur in winter than in summer.

2. High air and water radon levels are most commonly found in homes built over two-mica granite, although indoor air concentrations are low over other plutonic rocks.

3. Sands and gravels contribute significantly larger amounts of radon to indoor air than do less permeable soils, but the difference is apparent only in the winter. Overburden type is one

of the more important predictors of where indoor radon might exceed 4 pCi/L.

4. Mobile homes have the lowest average air radon levels.

5. Electrically heated homes have a significantly higher risk for elevated air radon levels than do homes heated by other sources.

6. Limited data suggest that stone or brick houses are most likely to have high air radon.

7. The presence of a basement did not significantly increase the risk of having indoor radon levels above 4 pCi/L.

Based on these findings, we recommend further air radon sampling of brick and stone houses, and that air radon screening protocols for houses with a drilled well in two-mica granite include testing at the level of the house where water use is the heaviest.

ACKNOWLEDGMENTS

This research was a cooperative effort involving the Maine Geological Survey, the Maine Medical Center, the University of Maine, and the Maine Department of Human Services. The work was begun in 1982 by E.M.L. as a master's thesis at the State University of New York, Stony Brook, and was continued at the Maine Geological Survey. Additional funding for this research was made possible by grants from the Maine State Legislature; the Guy Gannett Foundation; the American Cancer Society, Maine Division; the Maine Cancer Research and Education Foundation; the Maine Lung Association; and the Maine Medical Center Cancer Research Fund. We wish to gratefully acknowledge Steven Lawrence for his assistance in data collection; Walter A. Anderson, Maine State Geologist, for his support and consultation; and Roger Grimson, SUNY Stony Brook, for his assistance with the data analysis. Our special thanks go to Marc Loiselle, Maine Geological Survey, and to Steven Norton, Geology Department, University of Maine, for reviews of this work.

REFERENCES CITED

Åkerblom, G. V., and Wilson, C., 1981, Radon gas—A radiation hazard from radioactive bedrock and building materials: Bulletin of the International Association of Engineering Geology, no. 23, p. 51–61.

Breslow, N. E., and Day, N. E., 1980, Statistical methods in cancer research, Vol. I: The analysis of case-control studies: Lyon, France, International Agency for Research in Cancer, 338 p.

Buchli, R., and Burkart, W., 1989, Influence of subsoil geology and construction techniques on indoor air ^{222}Rn levels in 80 houses of the central Swiss Alps: Health Physics, v. 56, p. 423–429.

Centers for Disease Control, 1989, Radon exposure assessment—Connecticut: Morbidity and Mortality Weekly Report, v. 38, p. 713–715.

Cohen, B. L., and Gromicko, N., 1988, Variation of radon levels in U.S. homes with various factors: Journal of the Air Pollution Control Association, v. 38, p. 129–134.

Fleischer, R. L., Mogro-Campero, A., and Turner, L. G., 1983, Indoor radon levels in the northeastern United States: Effects of energy efficiency in homes: Health Physics, v. 45, p. 407–412.

George, A. C., and Breslin, A. J., 1980, The distribution of ambient radon and radon daughters in residential buildings in the New Jersey–New York area, in Natural Radiation Environment III, Vol. 2: Oak Ridge, Tennessee, National Technical Information Center, p. 1272–1292.

Grune, W. N., Higgins, F. B., and Smith, B. M., 1960, Natural radioactivity in groundwater supplies in Maine and New Hampshire: U.S. Public Health Service Contract No. 73551.

Guidotti, C. V., 1985, Generalized map of regional metamorphic zones, in Osberg, P. H., Hussey, A. M., II, and Boone, G. M., eds., Bedrock geologic map of Maine: Augusta, Maine Geological Survey, scale 1:1,584,000.

Hall, F. R., Boudette, E. L., and Olszewski, W. J., Jr., 1987, Geologic controls and radon occurrence in New England, in Graves, B., ed., Radon, radium and other radioactivity in ground water: Hydrogeologic impact and application to indoor airborne contamination: Chelsea, Michigan, Lewis Publishers, p. 15–29.

Hess, C. T., Norton, S. A., Brutsaert, W. F., Casparius, R. E., Coombs, E. G., and Hess, A. L., 1979, Radon-222 in potable water supplies in Maine: The geology, hydrology, physics and health effects: Orono, Land and Water Resources Center, University of Maine, 119 p.

Hess, C. T., Weiffenbach, C. V., and Norton, S. A., 1982, Variations of airborne and waterborne radon-222 in houses in Maine: Environment International, v. 8, p. 59–66.

Hess, C. T., Weiffenbach, C. V., and Norton, S. A., 1983, Environmental radon and cancer correlations in Maine: Health Physics, v. 45, p. 339–348.

Hess, C. T., Fleischer, R. L., and Turner, L. G., 1985, Field and laboratory tests of etched track detectors for ^{222}Rn: Summer-vs-winter variations and tightness effects in Maine houses: Health Physics, v. 49, p. 65–79.

Hoxie, D. C., 1966, A critical essay: Radon 222 determinations in ground waters: New Brunswick, New Jersey: Rutgers University (master's thesis), 24 p.

Kahlos, H., and Asikainen, M., 1980, Internal radiation doses from radioactivity of drinking water in Finland: Health Physics, v. 39, p. 108–111.

Kruskal, W. H., and Wallis, W. A., 1952, Use of ranks in one-criterion variance analysis: Journal of the American Statistical Association, v. 47, p. 583–621.

Mushrush, G. W., and Mose, D. G., 1988, Regional variation of indoor radon over three seasons: Environmental Toxicology and Chemistry, v. 7, p. 879–887.

Nazaroff, W. W., Moed, B. A., and Sextro, R. G., 1988, Soil as a source of indoor radon: Generation, migration and entry, in Nazaroff, W. W., and Nero, A., Jr., eds., Radon and its decay products in indoor air: New York, John Wiley & Sons, p. 57–112.

Osberg, P. H., Hussey, A. M., III, and Boone, G. M., compilers, 1985, Bedrock geologic map of Maine: Maine Geological Survey, scale 1:500,000.

Prichard, H. M., and Gesell, T. F., 1978, Rapid measurements of ^{222}Rn concentrations in water with a commercial liquid scintillation counter: Health Physics, v. 33, p. 577–581.

Radford, E. P., 1985, Potential health effects of indoor radon exposure: Environmental Health Perspectives, v. 62, p. 283.

Reimer, G. M., and Gundersen, L.C.S., 1989, A direct correlation among indoor Rn, soil gas Rn and geology in the Reading Prong near Boyertown, Pennsylvania: Health Physics, v. 57, p. 155–160.

Roessler, C. E., Roessler, G. S., and Bolch, W. E., 1983, Indoor radon progeny exposure in the Florida phosphate mining region: A review: Health Physics, v. 45, p. 389–396.

Thompson, W. B., and Borns, H. W., Jr., compilers, 1985, Surficial geologic map of Maine: Maine Geological Survey, scale 1:500,000.

U.S. Environmental Protection Agency, 1986, A citizen's guide to radon: What it is and what to do about it: Washington, D.C., U.S. Government Printing Office, OPA-86-004, p. 6.

Wilcoxon, F., 1945, Individual comparisons by ranking methods: Biometrics Bulletin, v. 1, p. 80–83.

Wilkening, M., and Wicke, A., 1986, Seasonal variation of indoor Rn at a location in the southwestern United States: Health Physics, v. 51, p. 427–436.

MANUSCRIPT ACCEPTED BY THE SOCIETY APRIL 6, 1992

Typeset by WESType Publishing Services, Inc., Boulder, Colorado
Printed in U.S.A. by Malloy Lithographing, Inc., Ann Arbor, Michigan